National 5
BIOLOGY
STUDENT BOOK

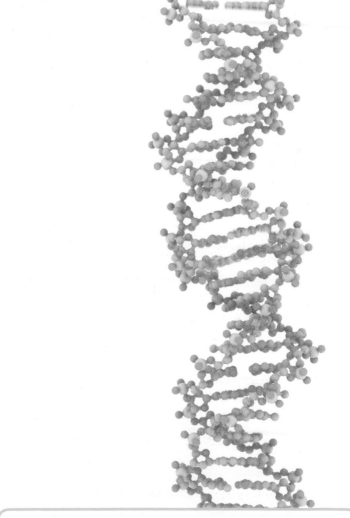

N5 BIOLOGY *STUDENT BOOK*

Claire Bocian • Dianne Forrest
Bryony Smith

© 2013 Leckie & Leckie Ltd

001/27052013

10 9 8 7 6 5 4 3

The authors assert their moral rights to be identified as the authors of this work.

ISBN 9780007504640

Published by
Leckie & Leckie Ltd
An imprint of HarperCollins*Publishers*
Westerhill Road, Bishopbriggs, Glasgow, G64 2QT
T: 0844 576 8126   F: 0844 576 8131
leckieandleckie@harpercollins.co.uk   www.leckieandleckie.co.uk

Special thanks to
Jouve (layout); Ink Tank (cover design); Paul Sensecall (proofread); Delphine Lawrance (copy-edit); Alistair Coats (proofread)

A CIP Catalogue record for this book is available from the British Library.

Acknowledgements
Whilst every effort has been made to trace the copyright holders, in cases where this has been unsuccessful, or if any have inadvertently been overlooked, the Publishers would gladly receive any information enabling them to rectify any error or omission at the first opportunity.

Leckie & Leckie would like to thank the following copyright holders for permission to reproduce their material:

Cover image © Lculig

Assessment questions Unit 1 A9, B7, B8, B9; Unit 2 A2, A4, A5, A9, A10, B4, B6; Unit 3 A10, B2, B4, B6 adapted or taken from past examination papers © Scottish Qualifications Authority. The solutions do not emanate from the SQA.

Unit 1 Impact: Dimarion; 1.1.1 Jupiterimages; 1.1.2 Duncan Smith; 1.1.11 THOMAS DEERINCK, NCMIR/SCIENCE PHOTO LIBRARY; 1.3.7 Roz Woodward; 1.3.24 MAURO FERMARIELLO/SCIENCE PHOTO LIBRARY; 1.3.25 SOUTHERN ILLINOIS UNIVERSITY/ SCIENCE PHOTO LIBRARY; 1.4.4 A. BARRINGTON BROWN/ SCIENCE PHOTO LIBRARY; 1.5.8 MARTYN F. CHILLMAID/ SCIENCE PHOTO LIBRARY; 1.6.1 PETER GARDINER/SCIENCE PHOTO LIBRARY; 1.7.3 CORDELIA MOLLOY/SCIENCE PHOTO LIBRARY; 1.7.9 SCIENCE PHOTO LIBRARY; 1.7.12 POWER AND SYRED/SCIENCE PHOTO LIBRARY; Unit 2 Impact: Jiri Vaclavek; 2.2.4 SCIENCE PHOTO LIBRARY; 2.2.5 J.C. REVY, ISM/SCIENCE PHOTO LIBRARY; 2.2.8 SINCLAIR STAMMERS/SCIENCE PHOTO LIBRARY; 2.2.9 DR COLIN CHUMBLEY/SCIENCE PHOTO LIBRARY; 2.2.13 MEHAU KULYK/SCIENCE PHOTO LIBRARY; 2.2.18 ALAIN POL, ISM/SCIENCE PHOTO LIBRARY; 2.2.21 Jupiterimages; 2.2.26; CNRI/SCIENCE PHOTO LIBRARY; 2.2.33 Jeffrey Hamilton; 2.2.36 George Doyle & Ciaran Griffin; 2.2.46 POWER AND SYRED/SCIENCE PHOTO LIBRARY; 2.2.52 DR JEREMY BURGESS/SCIENCE PHOTO LIBRARY; 2.2.56 Comstock Images; 2.2.57 Brand X Pictures; 2.6.1 Jupiterimages and Stockbyte; 2.2.62 White rose from George Doyle; 2.2.73 SCIENCE PHOTO LIBRARY; 2.2.75 TONY CAMACHO/SCIENCE PHOTO LIBRARY; 2.2.83 DR KEITH WHEELER/SCIENCE PHOTO LIBRARY; 2.2.87 POWER AND SYRED/SCIENCE PHOTO LIBRARY; 2.2.95 DR KEITH WHEELER/SCIENCE PHOTO LIBRARY; 2.2.105 EYE OF SCIENCE/ SCIENCE PHOTO LIBRARY; 2.2.115 SCIENCE PHOTO LIBRARY; 2.2.126 PROF. P. MOTTA/DEPT. OF ANATOMY/UNIVERSITY "LA SAPIENZA", ROME/SCIENCE PHOTO LIBRARY; 2.2.139 Jupiterimages; 2.2.140 Jupiterimages; Unit 3 Impact: SiberianLena; 3.3.2 Digital Vision; 3.3.4 Anup Shah; 3.3.33 NIGEL CATTLIN/ SCIENCE PHOTO LIBRARY; 3.3.46 PHILIPPE PSAILA/SCIENCE PHOTO LIBRARY; 3.3.54 SHEILA TERRY/SCIENCE PHOTO LIBRARY; 3.3.63 MICHAEL W. TWEEDIE/SCIENCE PHOTO LIBRARY; 3.3.65 DR. STANLEY FLEGLER, VISUALS UNLIMITED / SCIENCE PHOTO LIBRARY; 3.3.71 Mr Green / Shutterstock; 3.3.92 Comstock

UNIT 1 – CELL BIOLOGY

UNIT 2 – MULTICELLULAR ORGANISMS

UNIT 3 – LIFE ON EARTH

CONTENTS

Activities answers at   **www.leckieandleckie.co.uk/N5Biology**

End of Unit Assessment answers at   **www.leckieandleckie.co.uk/N5Biology**

Introduction

Welcome to the National 5 Biology Student Book!

This book covers all of the skills, knowledge and understanding included in the National 5 Course, following the published National 5 Course Support Notes and Assessment Specification.

It has been designed to help you pass National 5 Biology and is suitable for students who are currently following a National 5 course. Entry to the Course is at the discretion of a centre. However, we have assumed some prior skills and knowledge that would be gained from completing relevant Curriculum for Excellence experiences and outcomes and/or the National 4 Biology Course.

The book has three main units which are subdivided into chapters. Each chapter covers the mandatory knowledge with integrated skills. The book has a number of features to assist your learning.

Features

YOU SHOULD ALREADY KNOW:

Each chapter starts with a summary of the assumed prior knowledge. This should help to give you a background to the ideas that will be explored within the chapter.

You should already know:

- Plants use light energy to convert carbon dioxide and water into oxygen and glucose by a process known as photosynthesis.

LEARNING INTENTIONS:

Each chapter opens with a list of topics covered in the chapter, and gives you a good idea as to what you should be able to do when you have worked your way through the chapter.

Learning intentions

- Explain the need for transport systems in plants.

HINTS

Where appropriate, Hints are given to help give extra support and assist your learning.

> **🔍 Hint**
>
> The term 'nucleus' describes a single nucleus. If there is more than one nucleus, we say 'nuclei'.
>
> The term 'mitochondrion' describes a single mitochondrion. If there is more than one mitochondrion, we say 'mitochondria'.

MAKE THE LINK

There are two types of Make the Link featured in the book. Make the Link - Biology shows how one topic links to another unit in the National 5 course. Other Make the Links highlight links between subjects and help to show how the knowledge and skills gained from studying biology can be transferred to other areas.

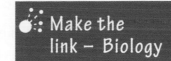

Make the link – Biology

Within Unit 1 you will learn more about what happens in each of the cell structures. The role of ribosomes in protein synthesis (Chapter 4) and the role of mitochondria in aerobic respiration (Chapter 8) will be examined in greater detail.

BIOLOGY IN CONTEXT

Throughout the book, examples are given that show the real-life applications of the topics in the course.

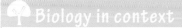

Biology in context

The majority of the human population have a chromosome complement of 46 chromosomes. However, some children are born with the genetic condition Down's syndrome.

ACTIVITIES

These are sets of questions covering knowledge and skills and active learning tasks to work on individually, with a partner or in a group.

GO! Activities

Activity 1.1 Working individually

(a) Make drawings of the ultrastructure of each of the four different cell types. Add arrows to identify each of the different structures in each of the cells. Using a separate piece of paper, cut out a set of labels and practise identifying the different structures in each of your drawings.

SUCCESS CRITERIA

Each chapter closes with a summary of learning statements showing what you should be able to do when you complete the chapter. You should use them to help identify where you are with your learning and the next steps needed to make any improvements.

I can:

• Name and identify the following structures found in the cell ultrastructure of an animal cell: nucleus, cell membrane, cytoplasm, mitochondria and ribosomes.

ASSESSMENT

End of unit assessments are provided at the end of each unit.

These take a similar format to the question papers you will sit in the National 5 exam.

ANSWERS

Answers to knowledge & understanding activities and the End of Unit Assessments are provided online at www.leckieandleckie.co.uk/N5Biology.

Course award

To gain a course award you must pass all three units, as well as the course assessment.

Unit assessment

The Outcomes and Assessment Standards are as follows:

Outcomes	Assessment standards	
Outcome 1 The candidate will: **Apply skills of scientific inquiry and draw on knowledge and understanding of the key areas of this Unit to carry out an experiment/practical investigation by:**	1.1 Planning an experiment/practical investigation	
	1.2 Following procedures safely	
	1.3 Making and recording observations/measurements accurately	
	1.4 Presenting results in an appropriate format	
	1.5 Drawing valid conclusions	
	1.6 Evaluating experimental procedures	
Outcome 2 The candidate will: **Draw on knowledge and understanding of the key areas of this Unit and apply scientific skills by:**	2.1 Making accurate statements	Unit 1
		Unit 2
		Unit 3
	2.2 Describing an application	
	2.3 Describing a biological issue in terms of the effect on the environment/society	
	2.4 Solving problems	Predicting
		Selecting
		Processing
		Analysing

Evidence for Outcome 1 and 2.2, 2.3 and 2.4 of Outcome 2 need only be demonstrated once across the course. Evidence for 2.1 must be demonstrated across all three units.

Course assessment

Course assessment provides the basis for grading attainment in the course award.

Course assessment structure

COMPONENT 1 – THE QUESTION PAPER

The question paper will have two Sections.

Section 1, titled 'Objective Test', will have 20 marks.

Section 2, titled 'Paper 2', will contain restricted and extended response questions and will have 60 marks.

Total of 80 marks (80% of the total mark).

Marks will be distributed approximately proportionately across the units.

The majority of the marks will be awarded for applying knowledge and understanding. The other marks will be awarded for applying scientific inquiry, scientific analytical thinking and problem-solving skills.

COMPONENT 2 – THE ASSIGNMENT

The purpose of the assignment is to allow the learner to carry out an in-depth study of a biology topic. The topic will be chosen by the learner, who will investigate/research the underlying biology and the impact on the environment/society.

The assignment will assess the application of skills of scientific inquiry and related biology knowledge and understanding.

Candidates should only start the assignment when they have sufficient knowledge of the course units and skills to undertake it. It is recommended that no more than eight hours is spent on the whole assignment.

The assignment is split into a research stage and a communication stage, and candidates may produce their report over a period of time. If the report is done over a number of sessions, then the assessor must retain the candidate's work between sessions. Following completion of the report there should be no re-drafting.

As a guide, evidence that meets the requirements of this component of course assessment should be 500–800 words, excluding tables, charts and diagrams. There is no penalty for being outwith this range.

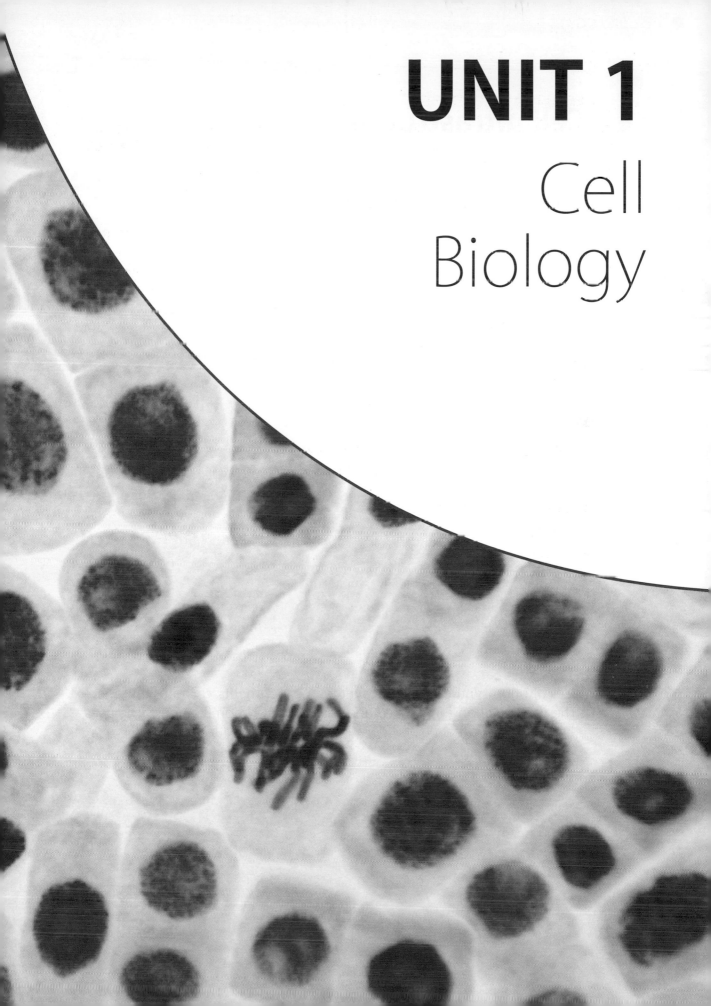

UNIT 1
Cell
Biology

1 Cell ultrastructure

You should already know:

- Cells are the basic units of all living things.
- Animal and plant cells have a nucleus, cell membrane and cytoplasm.
- Plant cells also have a cell wall, sap vacuole and chloroplasts.
- The cell membrane controls what enters and leaves the cell.
- The nucleus contains genetic information and controls all cell activities.
- The cytoplasm is the site of chemical reactions.
- The cell wall is freely permeable and is involved in support of the plant cell.
- The sap vacuole helps to keep the shape of the cell.
- The chloroplasts are the site of photosynthesis.

Learning intentions

- Name the structures found in the ultrastructure of an animal cell.
- Identify in a diagram the structures found in the ultrastructure of an animal cell.
- State the functions of the structures found in the ultrastructure of an animal cell.
- Name the structures found in the ultrastructure of a plant cell.
- Identify in a diagram the structures found in the ultrastructure of a plant cell.
- State the functions of the structures found in the ultrastructure of a plant cell.
- Name the structures found in the ultrastructure of a fungal cell.
- Identify in a diagram the structures found in the ultrastructure of a fungal cell.
- State the functions of the structures found in the ultrastructure of a fungal cell.
- State that a fungal cell has a different cell wall structure to a plant cell.
- Name the structures found in the ultrastructure of a bacterial cell.
- Identify in a diagram the structures found in the ultrastructure of a bacterial cell.
- State the functions of the structures found in the ultrastructure of a bacterial cell.
- State that a bacterial cell has a different cell wall structure to both plant and yeast cells.

Basic cell structure

The basic structures of animal and plant cells can be viewed using a light microscope (fig 1.1). Simple biological line drawings can be made of these cells.

Cheek epithelial cells are a good example of basic animal cells. A swab can be taken from the inside of the cheek and examined under the microscope. Biological stains are used on cells to allow the different cell structures to be clearly viewed. Using a light microscope allows some of the organelles that are present in animal cells to be seen.

Figure 1.2 shows the basic structures visible in an animal cell using a light microscope, e.g. cheek epithelial cells.

The names of each structure found in a basic animal cell are shown in the simple illustration below (fig 1.3).

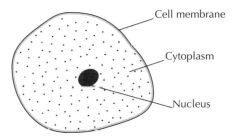

Fig 1.3 *Basic animal cell*

Cells from onion, rhubarb and elodea (Canadian pondweed) are just some examples of plant cells that can be viewed under the light microscope.

Onion epidermal cells can be stained with iodine and viewed using a light microscope (fig 1.4).

Like an animal cell, a plant cell has a nucleus, cytoplasm and cell membrane. Also visible in figure 1.4 is the cell wall. Chloroplasts are only visible when green parts of a plant are viewed under the microscope.

In figure 1.5, the chloroplasts are clearly visible when elodea is viewed under the microscope.

The line drawing (fig 1.6) shows the names of each structure found in a plant cell.

Fig 1.1 *Using a light microscope*

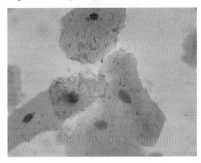

Fig 1.2 *Cheek epithelial cells viewed under a light microscope*

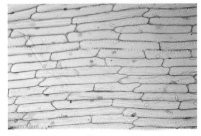

Fig 1.4 *Plant cells viewed under a light microscope*

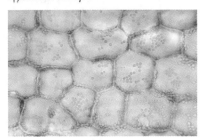

Fig 1.5 *Elodea (Canadian pondweed) cells under a light microscope*

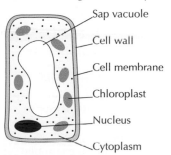

Fig 1.6 *The parts of a plant cell*

15

Fig 1.7 *Light microscope*

Fig 1.8 *Electron microscope*

🔍 **Hint**

The term 'nucleus' describes a single nucleus. If there is more than one nucleus, we say 'nuclei'.

The term 'mitochondrion' describes a single mitochondrion. If there is more than one mitochondrion, we say 'mitochondria'.

⚫∴ Make the link – Biology

Within Unit 1 you will learn more about what happens in each of the cell structures. The role of ribosomes in protein synthesis (Chapter 4) and the role of mitochondria in aerobic respiration (Chapter 8) will be examined in greater detail.

As a light microscope (fig 1.7) has limited magnification, an electron microscope (fig 1.8) is used to view cells in greater detail. With this microscope other organelles present in the cytoplasm become visible. What can be seen at cell level is described as the **cell ultrastructure**.

Ultrastructure of an animal cell

In addition to the **nucleus**, **cell membrane** and **cytoplasm,** other structures can be seen using the electron microscope.

The illustration in figure 1.9 shows the ultrastructure of an animal cell as revealed by an electron microscope. In addition to the nucleus, cytoplasm and cell membrane, there are two other organelles. The **mitochondrion** is the site of aerobic respiration, and the **ribosomes** are where proteins are synthesised.

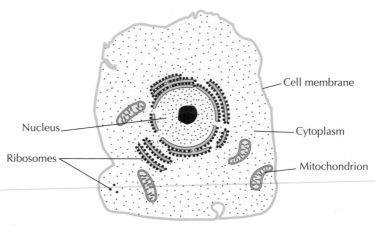

Fig 1.9 *Animal cell ultrastructure*

Cells can be thought of as mini factories with jobs being carried out in specialised areas. Each structure is designed in a way that suits its function. Specific organelles carry out their function to ensure the survival of the cell.

Ultrastructure of a plant cell

The structures normally visible in a plant cell under the light microscope are: the nucleus, cell membrane, cytoplasm, **cell wall**, **sap vacuole** and **chloroplasts**. Using an electron microscope allows you to see the chloroplast in more detail. Ribosomes and mitochondria are also visible and have the same function as in animal cells.

The illustration below shows the organelles present in a plant cell (fig 1.10).

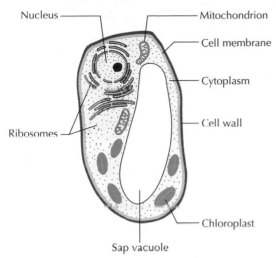

Nucleus — Mitochondrion — Cell membrane — Cytoplasm — Cell wall — Chloroplast — Sap vacuole — Ribosomes

Fig 1.10 *The structure of a plant cell*

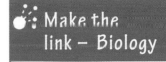

Make the link – Biology

Within Unit 1 you will learn more about the process of photosynthesis that takes place inside chloroplasts (Chapter 7).

Ultrastructure of a fungal cell

Yeast is an example of a unicellular fungus. The structure of a yeast cell is very similar to an animal cell. It has a nucleus, cell membrane and cytoplasm (fig 1.11). The only difference between the animal and yeast cell is the presence of a vacuole and cell wall in the yeast cell. The structures perform the same functions in animal, plant and yeast cells. The yeast cell has structures common to both animal and plant cells. The cell walls in plant cells and yeast cells have a different composition.

There are a variety of different types of yeast. Different strains of yeast are used in the baking and brewing industries.

Fig 1.11 *Yeast cells in colour under a high-power microscope*

The illustration in figure 1.12 shows the organelles present in a yeast cell.

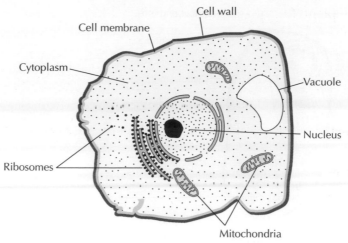

Fig 1.12 *The organelles present in a yeast cell*

Ultrastructure of a bacterial cell

There are many different types of bacteria. Like the other cell types, the bacterial cell has a cell membrane, cell wall, cytoplasm and ribosomes.

The most obvious difference between a bacterial cell and the other cell types is the lack of a nucleus. The bacterial cell contains DNA as well as circular **plasmids** that also contain DNA. The cell wall in the bacterial cell is again different to those in plant and yeast cells. Although all three cell types (plant, fungal and bacterial cells) have a cell wall, the chemical composition of all three is different.

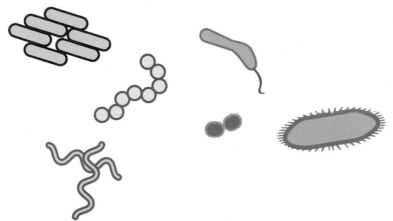

Fig 1.13 *The different forms bacterial cells can take*

Figure 1.13 shows that bacterial cells come in many different shapes and sizes. Some bacteria are harmful, while others are useful.

The illustration in figure 1.14 shows the structures present in a typical bacterial cell.

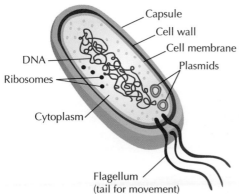

Fig 1.14 *The structures present in a typical bacterial cell*

Cell functions

The table below summarises the functions of all structures found in animal, plant, fungal and bacterial cells.

Structure	Function
Nucleus	Contains genetic information and controls all cell activities
Cell membrane	Selectively permeable; controls what enters and leaves the cell
Cytoplasm	Site of chemical reactions
Mitochondria (-ion)	Site of aerobic respiration (energy production)
Ribosome	Site of protein synthesis
Cell wall	Freely permeable; involved in support of the plant
Sap vacuole	Helps keep the shape of the cell
Chloroplast	Site of photosynthesis
Plasmid	Circular piece of DNA found in bacterial cells

Cell sizes

Before the invention of the microscope, people had never seen cells. As technology advances, so do microscopes. This allows us to see the complexity of cells.

The size of an individual cell can be calculated based on the number of cells that are visible in the field of view. To calculate the size of an individual cell, we need to know the total magnification of the microscope. The total magnification is calculated by multiplying the eyepiece lens magnification by the objective lens magnification.

For example, if the eyepiece lens used was ×10, and the objective lens was ×4, the total magnification would be ×40.

As cells are microscopic they need to be measured in units smaller than millimetres, so they are measured in micrometres (also referred to as microns) and nanometres. 1 micrometre (μm) is equal to one thousandth of a millimetre. 1 nanometre is one thousandth of a micrometre. Using a light microscope we can see structures that can be measured in micrometres. With an electron microscope, structures measured in nanometres are visible. Figure 1.15 helps to put these sizes into context.

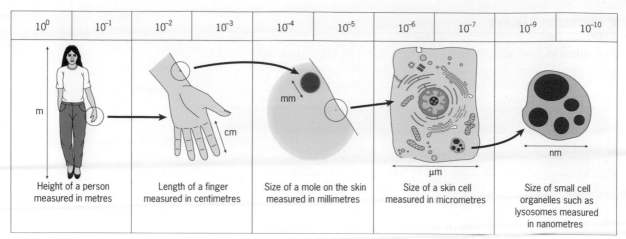

10^0	10^{-1}	10^{-2}	10^{-3}	10^{-4}	10^{-5}	10^{-6}	10^{-7}	10^{-9}	10^{-10}

Height of a person measured in metres

Length of a finger measured in centimetres

Size of a mole on the skin measured in millimetres

Size of a skin cell measured in micrometres

Size of small cell organelles such as lysosomes measured in nanometres

Fig 1.15 *The relationship between units of measurement*

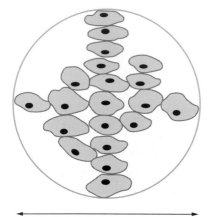

Field of view = 2 mm

Fig 1.16 *Measuring cell length using a microscope*

In this example, the field of view is 2 mm in diameter.

To work out the length of an individual cell, the number of cells across the length of the field of view must be counted.

There are five cells across the field of view

$$\frac{\text{Field of view}}{\text{Number of cells}} \times 1000 = \text{length of each cell.}$$

So, $2 \div 5 = 0.4$ mm. Therefore the length of each cell is equal to 400 μm (fig 1.16).

🔵 GO! Activities

Activity 1.1 Working individually

(a) Make drawings of the ultrastructure of each of the four different cell types. Add arrows to identify each of the different structures in each of the cells. Using a separate piece of paper, cut out a set of labels and practise identifying the different structures in each of your drawings.

(b) On paper, draw a structure and function table. Complete the structures and functions using the table on page 19. Repeat the process using a mini whiteboard. This time try to complete the table without using your notes.

(c) Using a mini whiteboard, copy the table below. Add ticks to show which structures are found in each of the cell types.

Structures	Animal	Plant	Fungal	Bacterial
Cell membrane				
Cytoplasm				
Nucleus				
Cell wall				
Sap vacuole				
Chloroplast				
Mitochondria				
Ribosome				
Plasmid				

Activity 1.2 Working in pairs

(a) Each person should make nine small cards. One person writes down the names of the nine different cell structures. The other person writes down the functions of the nine structures. Work together to match up each structure with the correct function.

(b) On a sheet of paper, each person should copy and complete the table below. Work together, using the formula to calculate the total magnification based on the same eyepiece being used each time.

Total magnification = eyepiece-lens magnification × objective-lens magnification

Eyepiece-lens magnification	Objective-lens magnification	Total magnification
× 10	× 4	
	× 10	
	× 40	

(c) Now make your own diagrams of a 2 mm diameter field of view down a microscope. Insert between 2 and 20 cells then calculate the average length of a cell in microns. Show all of your calculations. Swap diagrams with a partner to check each other's calculations.

Activity 1.3 Working in groups

From memory, produce 4 different mini-posters to show the ultrastructure of each of the cell types – animal, plant, fungal and bacterial. Label all of the cell structures and describe their functions.

I can:

- Name and identify the following structures found in the cell ultrastructure of an animal cell: nucleus, cell membrane, cytoplasm, mitochondria and ribosomes.

- State the functions of the structures found in the ultrastructure of an animal cell.

- Name and identify the following structures found in the cell ultrastructure of a plant cell: nucleus, cell membrane, cytoplasm, cell wall, sap vacuole, chloroplast, mitochondria and ribosomes.

- State the functions of the structures found in the ultrastructure of a plant cell.

- Name and identify the following structures found in the cell ultrastructure of a fungal cell: nucleus, cell membrane, cytoplasm, cell wall, vacuole, mitochondria and ribosomes.

- State the functions of the structures found in the ultrastructure of a fungal cell.

- State that a fungal cell has a cell wall, but it is different in structure to that of a plant cell wall.

- Name and identify the following structures found in the cell ultrastructure of a bacterial cell: cell membrane, cytoplasm, cell wall, plasmids and ribosomes.

- State the functions of the structures found in the ultrastructure of a bacterial cell.

- State that a bacterial cell has a cell wall, but it is different in structure to that of a plant cell wall or fungal cell wall.

- Calculate total magnification using the formula:

  total magnification = eyepiece lens magnification × objective lens magnification

- Calculate the length and breadth of individual cells when given the diameter of the field of view.

2 Transport across cell membranes

The structure of the cell membrane

The cell membrane separates the cell contents from its external environment. It is made of **protein** and **lipid** molecules. The fluid mosaic model is commonly used to describe how proteins and lipids are arranged within the cell membrane.

The lipid molecules are found in a double or bi-layer that is in constant motion – this is the 'fluid' part. The flexibility of this double layer allows the cell to change in shape without causing damage to the cell.

The protein molecules are found in a patchy arrangement, spread throughout the lipid molecules – this gives the mosaic pattern. Some proteins pass through the width of the membrane and contain a channel, forming a pore. Other proteins are only partly embedded, and some lie on the surface (fig 2.1). The proteins have different functions.

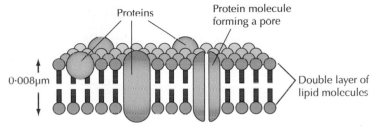

Fig 2.1 *The structure of the cell membrane*

The channel-forming protein pores allow certain substances to enter and leave the cell. Owing to this property, cell membranes are described as being **selectively permeable**. Small soluble molecules can pass easily though the membrane. Large insoluble molecules must first be broken down before they can pass into or out of cells.

> **🔍 Hint**
>
> As cells are microscopic they are measured in micrometres (μm).
> 1 micrometre = 1 thousandth of a millimetre.
> The cell membrane is between 7 and 10 μm.

🔵 Activities

Activity 2.1 Working individually

Using a mini whiteboard, practise drawing and labelling the structure of the membrane. Make sure that you can correctly identify the proteins and lipids from memory.

Activity 2.2 Working in pairs

Described below is an experiment that is carried out to investigate the structure of the cell membrane. With your partner, read the information carefully and complete the task.

If the membrane is made of lipids or proteins, damage to either one of them should cause leakage of the cell contents. Red cabbage can be used to investigate the structure of the cell membrane. The vegetable is dark purple in colour, so any damage to the membrane results in the purple liquid leaking from the tissue. Proteins can be damaged at high temperatures, and lipids can be dissolved by alcohol.

The following experiment was carried out to investigate the composition of the membrane. 40 discs of red cabbage were cut and then rinsed in water until the water ran clear. They were then placed in test tubes under the following conditions:

Test tube 1: 10 cm³ of water and 10 discs of red cabbage left at room temperature for 15 minutes.

(Continued)

Test tube 2: 10 cm³ of water and 10 discs of red cabbage placed in a water bath at 90°C for 15 minutes.

Test tube 3: 10 cm³ of alcohol and 10 discs of red cabbage left at room temperature for 20 minutes

Test tube 4: 10 cm³ of alcohol and 10 discs of red cabbage placed in a water bath at 90°C for 15 minutes.

Use figs 2.2a and b to make notes and draw conclusions on the structure of the cell membrane. Assess each test tube individually using the list below to help you to focus on each experiment in turn.

Fig 2.2a *Discs of red cabbage at the start of the experiment*

Examine – The before and after photographs for each test tube individually. Note that the cabbage has not disappeared; it has sunk to the bottom of each test tube. Examine the colour changes.

Observe – What do you notice about the colour of the water?

Discuss – Why has this change taken place? What condition (temperature/alcohol) caused the change? What part of the membrane (protein/lipid) has been affected?

Fig 2.2b *Discs of red cabbage following various treatments*

Conclude – What effect did the condition have on the cell membrane? Why did this happen?

Movement of substances across the cell membrane

The movement of substances into and out of cells is essential for all living organisms. Movement of substances through the cell membrane occurs by three different transport methods:

(1) Diffusion

(2) Osmosis

(3) Active transport

Different concentrations of substances exist between cells and their environment. In order to understand the movement of substances into and out of cells, we need to understand the term 'concentration gradient'.

A **concentration gradient** can be described as a difference in concentration of a substance between two solutions or between two cells or cell/tissue and a solution. Within any liquid or gas, a concentration gradient can exist. This happens when there is an uneven distribution of molecules (fig 2.3).

In figure 2.3, the molecules move from where they are in a higher concentration to where they are in a lower concentration. This happens until the concentration of molecules is equal. The movement of these molecules is described in the process of **diffusion**.

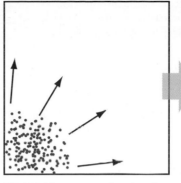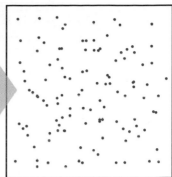

Fig 2.3 *Movement of molecules*

Passive transport

Passive transport is the movement of substances along a concentration gradient and does not require energy. Small molecules can pass easily through a cell membrane by **passive transport.**

Diffusion

Diffusion is a process by which some molecules move into and out of cells. **Diffusion** is defined as **the movement of molecules of a substance from an area of high concentration to an area of low concentration until they are evenly spread**. It is a passive process and does not require energy.

The importance of diffusion

Diffusion is essential to all living things. All cells need to take in food (e.g. glucose and amino acids) and oxygen. These can be used in respiration to provide energy needed by the cell to carry out all of its activities. Waste substances such as carbon dioxide can be removed from cells by diffusion.

Diffusion in action

Diffusion can only occur when there is a difference in concentration of a substance. The difference in concentration of a substance creates the concentration gradient. Molecules of a substance diffuse down a concentration gradient so diffusion requires no energy. Diffusion is an example of passive transport.

Diffusion can be easily demonstrated in a simple experiment using potassium permanganate crystals. A crystal of potassium permanganate can be placed in a beaker of still water. The crystal has a deep purple colour and will dissolve in the water over a period of time. Eventually the water will be the same colour throughout (fig 2.4).

⚫⁚ Make the link – Biology

Diffusion and osmosis are both examples of passive transport. They are examined in greater detail later in this chapter.

⚫⁚ Make the link – Biology

In Chapter 8 of this Unit you will learn about respiration in more detail. Diffusion is essential in respiration to allow the exchange of substances between cells and their environment.

In Unit 2, you will learn about digestion in the small intestine (p. **206**) and gas exchange in the lungs (p. **201**). Diffusion is essential in both of these processes.

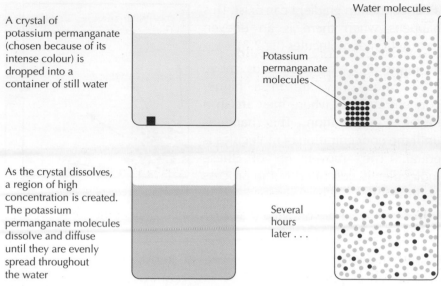

A crystal of potassium permanganate (chosen because of its intense colour) is dropped into a container of still water

Water molecules

Potassium permanganate molecules

As the crystal dissolves, a region of high concentration is created. The potassium permanganate molecules dissolve and diffuse until they are evenly spread throughout the water

Several hours later . . .

Fig 2.4 *Diffusion of potassium permanganate*

At the start of the experiment, a concentration gradient exists between the crystal and the water. The molecules of the crystal are in a high concentration – they are clustered together. Over time they spread throughout the water, moving down the concentration gradient until the concentration is equal. By the end of the experiment, the water is the same colour throughout. The molecules have moved passively by diffusion.

🌳 Biology in context

Dissolving sugar in a cup of tea is an everyday example of diffusion. Even if you use a sugar cube and don't stir the sugar, it will still dissolve in the water, giving a sugary taste to every sip. This is diffusion in action. The sugar molecules move down the concentration gradient from where the sugar is highly concentrated to where there is a low concentration until they are evenly spread (fig 2.5).

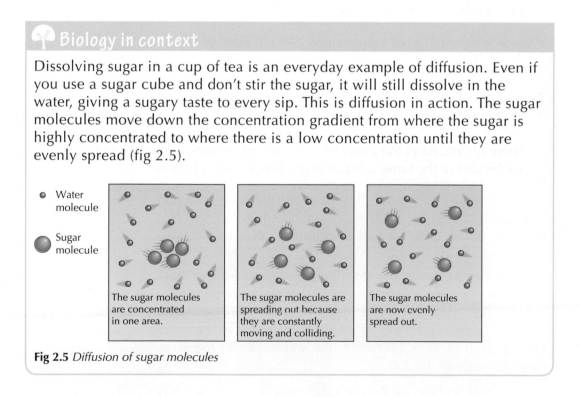

- Water molecule
- Sugar molecule

The sugar molecules are concentrated in one area.

The sugar molecules are spreading out because they are constantly moving and colliding.

The sugar molecules are now evenly spread out.

Fig 2.5 *Diffusion of sugar molecules*

ⒼⓄⒾ Activities

Activity 2.3 Working individually

(a) Write down full answers to the following questions:

1. Define the term 'passive transport'.
2. Define the term 'diffusion'.
3. Describe the relationship between diffusion and passive transport.
4. Using named substances, explain how the process of diffusion happens across cell membranes.

(b) Make a simple drawing of a cup of tea. Repeat the same drawing two more times.

In the first mug, add a sugar cube, as shown in fig 2.6, to show the particle arrangement of a solid.

Using small circles to represent sugar particles, show what happens to the concentration of sugar in the cup over time. The third cup should show what happens once diffusion is complete.

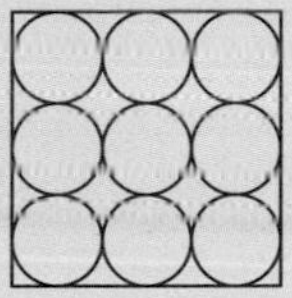

Fig 2.6 *Particle arrangement of a sugar cube*

Activity 2.4 Working individually or in pairs

Make a simple diagram of a plant or animal cell. You should remember this from Chapter 1 (to practise your knowledge of cell ultrastructure make a more detailed drawing).

- Show the movement of the following substances: glucose, amino acids (both are examples of dissolved foods), oxygen and carbon dioxide.
- Create a key using different colours and/or shapes for each substance.
- Think carefully about which substances move into the cell and which substances move out.
- All of these substances move by the passive process of diffusion.
- Make sure that you show the substances moving from a high concentration to a low concentration. To show a higher concentration, make sure there are more molecules of the substance in comparison with the number of molecules of the same substance at a lower concentration.

Osmosis

The diffusion of water is called **osmosis**. This is defined as **the movement of water molecules from an area of high water concentration to an area of low water concentration** through a selectively permeable membrane (fig 2.7). Water molecules pass easily through a selectively permeable membrane, as they are very small.

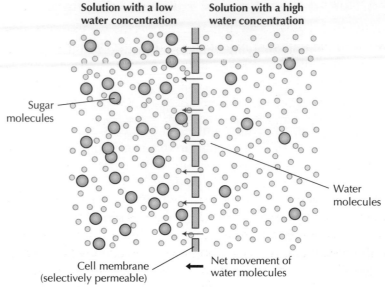

Fig 2.7 *Movement of water through a selectively permeable membrane*

The water molecules can pass freely through the membrane between the two solutions. Overall, there is a greater movement of water molecules from the solution with the higher water concentration to the solution with the lower water concentration. The water moves from where it is in a higher concentration to where it is in a lower concentration.

The effects of osmosis on cells

Different concentrations of solutions affect cells in different ways. This can be demonstrated by some simple experiments using model cells.

Visking tubing is selectively permeable and can be used to represent the cell membrane.

Two Visking tubing bags are set up as model cells. Both contain a weak sugar solution. One is placed into a beaker containing a solution of pure water (Beaker A) and the other into a beaker containing a strong sugar solution (Beaker B) (fig 2.8).

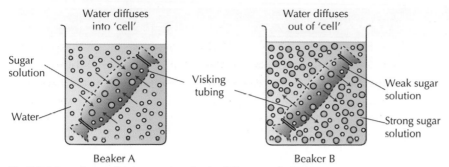

Fig 2.8 *Movement of water molecules in different solutions*

From figure 2.8 we can see that in beaker A the water molecules move from the region of high water concentration in the beaker of pure water into the visking tubing bag where there is a lower water concentration. In beaker B the water molecules move out of the visking tubing bag from a region of high water concentration into the strong sugar solution where there is a lower water concentration.

Osmosis in plant cells

Plant cells contain a weak sugar solution inside their sap vacuole. When a plant cell is placed into pure water, it swells but does not burst. This is due to the strong cell wall. When water enters the cell, the membrane is pushed up against the cell wall causing it to bulge. This can be seen using a microscope. Plant cells viewed under the microscope are described as being **turgid** (fig 2.9).

Plant cells placed into a concentrated sugar solution with a low water concentration lose water to the solution. Water moves out from a region of higher water concentration inside the cell to a region of lower water concentration outside the cell. This causes the membrane to draw away from the cell wall. The rigid structure of the cell wall prevents the plant cell changing in shape. When viewed under the microscope, these cells are described as being **plasmolysed** (fig 2.10).

Osmosis in animal cells

Animal cells, like red blood cells, contain a weak salt solution. When a red blood cell is placed into a solution of pure water, a concentration gradient is created. Water moves from the higher water concentration outside the cell to the lower concentration inside the cell. As an animal cell has no cell wall to give it protection, the blood cell swells with water and eventually **bursts** (fig 2.11).

If a red blood cell is placed into a solution with a lower water concentration than the cell, water will leave the cell and move into the solution. The water molecules move from a higher water concentration inside the cell to the lower water concentration outside the cell. When this happens the cell loses water, making it **shrink** (fig 2.12).

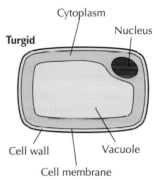

Fig 2.9 *A turgid plant cell*

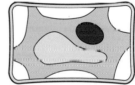

Plasmolysed

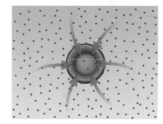

Fig 2.10 *A plasmolysed plant cell*

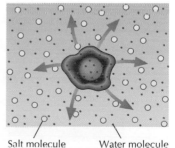

Fig 2.11 *Water moving into a red blood cell by osmosis*

Salt molecule   Water molecule

Fig 2.12 *Water moving out of a red blood cell by osmosis*

GO! Activity

Activity 2.5 Working individually

(a) Write down full answers to the following questions:

1. Define the term 'osmosis'.
2. Explain how the process of osmosis moves water across cell membranes.
3. Describe the osmotic effects of the following:
   - transferring animal cells from a weak salt solution to a strong salt solution
   - transferring animal cells from a weak salt solution to a solution of pure water
   - transferring plant cells from a weak sugar solution to a strong sugar solution
   - transferring plant cells from a weak sugar solution to a solution of pure water

(b) Read the information below then complete the task.

The experiment described below was set up to investigate the effects of osmosis on plant tissue (fig 2.13).

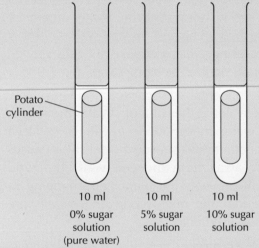

Potato cylinder

| 10 ml | 10 ml | 10 ml |
| 0% sugar solution (pure water) | 5% sugar solution | 10% sugar solution |

Fig 2.13 *Experiment to investigate the effects of osmosis on plant tissue*

1. Three boiling tubes were set up; 1 with 10 ml of pure water, 1 with 10 ml of 5% sugar solution and 1 with 10 ml of 10% sugar solution.
2. Three cylinders of potato were cut and blotted to remove excess water.
3. The mass of each cylinder was recorded before being placed into the different solutions.
4. The experiment was left set up for 2 hours.
5. The cylinders of potato were removed from the boiling tubes, blotted and re-weighed.

6. The mass of the potato cylinders was recorded.

7. The change in mass was calculated (difference between initial and final mass) and using the following calculation, the percentage change in mass was obtained.

$$\frac{\text{Change in mass}}{\text{initial mass}} \times 100 = \text{percentage change in mass}$$

The entire experiment was repeated a further two times and the results averaged to improve their reliability.

The results are shown in the table below.

Concentration of sugar solution (%)	Initial mass (g)	Final mass (g)	Change in mass (g)	% change in mass	Average change in mass (%)
	2.0	2.3	0.3	15	
0	2.5	2.6	0.1	4	+9.1
	2.4	2.6	0.2	8.3	
	2.3	2.2	−0.1	−4.3	
5	2.2	2.1	−0.1	−4.5	−6.9
	2.5	2.2	−0.3	−12	
	2.2	1.9	−0.3	−13.6	
10	2.4	1.9	−0.5	−20.8	−15.5
	2.5	2.2	−0.3	−12	

Using the results, the following graph was drawn (fig 2.14):

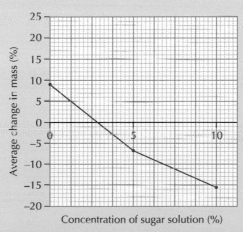

Fig 2.14 *Line graph of results from osmosis experiment*

From the results, it can be concluded that as the concentration of the sugar solution is increased, the average percentage change in mass decreases.

Complete the following tasks using the information from the osmosis experiment described.

1. The independent variable is the variable altered by the experimenter. Name the independent variable.

2. The dependent variable is the variable that changes as a result of altering the independent variable. Name the dependent variable.

3. Make a list of the variables that would need to be controlled in the experiment.

4. What was done in the experiment to ensure that the results obtained were reliable?

5. A similar experiment was repeated with different concentrations of sugar solutions. Again the experiment was repeated three times at each concentration.

   The following results were obtained:

Concentration of sugar solution (%)	Initial mass (g)	Final mass (g)	Change in mass (g)	% change in mass	Average change in mass (%)
	2.0	2.1	0.1	5	
0.2	2.2	2.3	0.1		
	2.4	2.5	0.1	4.2	
	2.1	2.0	−0.1		
0.4	2.2	1.9	−0.2	−9.1	−6.0
	2.4	2.3	−0.1		
	2.3	2.1	−0.2	−8.7	
0.6	2.0	1.6	−0.4	−20	
	2.4	2.1	−0.3	−12.5	

(a) Copy the table and complete it using the calculation detailed earlier in the previous question 7 on page 33.

(b) Using your completed table, draw a graph to show the average change in mass (%) at each of the different sugar concentrations.

Active transport

Active transport is **the movement of molecules and ions from a region of low concentration to a region of high concentration**. This is the opposite of diffusion. Diffusion is a passive process so does not require any energy. **Active transport** has to work **against the concentration gradient**, which **requires energy**.

The membrane proteins act like pumps to move molecules and ions across the membrane. As this goes against the concentration gradient, the proteins use energy.

In nerve cells, sodium and potassium are actively transported across the cell membrane. The protein molecules in the membrane act as carriers to transport these substances. Iodine is also transported this way in seaweed.

Figure 2.15 shows molecules moving from a region of lower concentration to a region of higher concentration, with a protein acting as a pump.

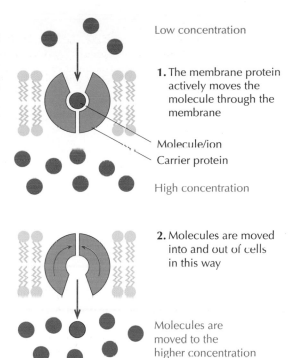

Low concentration

1. The membrane protein actively moves the molecule through the membrane

Molecule/ion
Carrier protein

High concentration

2. Molecules are moved into and out of cells in this way

Molecules are moved to the higher concentration

Fig 2.15 *Active transport of molecules*

🔵 GO! Activities

Activity 2.6 Working individually

Write down full answers to the following questions:

(a) Define the term 'active transport'.

(b) Explain how active transport of molecules occurs across cell membranes.

Activity 2.7 Working in pairs

Draw a diagram to show the structure of the cell membrane. Include at least four membrane proteins with channels. Label the lipids and proteins. Split your diagram in half. On one side of the diagram show examples of passive transport; on the other show active transport. Your diagram should include examples of diffusion, osmosis and active transport. Use different shapes and colours to represent the molecules that would move across the cell membrane. Add as much detail as possible to show your understanding of these topics.

Activity 2.8 Working in pairs

Make a set of revision cards on this topic. Use coloured paper or different colours of pen. On one side of the card draw a picture or write the word that you want to learn. On the other side write the definition. For example, on one side write the word 'Diffusion'; on the other side write what the word means. Use the cards to test each other.

I can:

- State that the cell membrane is made of lipids and proteins.

- Identify lipids and proteins on a diagram of the cell membrane.

- State that the cell membrane is selectively permeable.

- Explain that the membrane proteins have channels that allow substances to enter and leave the cell.

- Explain that the cell membrane is selectively permeable.

- State that passive transport is the movement of a substance down a concentration gradient and does not require energy.

- State that different concentrations of substances exist between cells and their environment.

- State that diffusion is the movement of a substance from an area of high concentration to an area of low concentration until evenly spread.

- Describe diffusion in terms of concentration gradients.

- State that diffusion is an example of passive transport.

- Name glucose, carbon dioxide, oxygen and amino acids as examples of substances that diffuse across cell membranes.

- Explain the importance of diffusion to organisms as being the means by which substances enter and leave cells by movement down the concentration gradient.

- Identify osmosis as a 'special case' of the diffusion of water.

- State that osmosis is the movement of water from an area of high concentration to an area of low concentration, down a concentration gradient, and across a selectively permeable membrane.

- Explain that animal cells can burst or shrink, and plant cells can become turgid or plasmolysed in different solutions.

- State that active transport is the movement of a substance against a concentration gradient and requires energy.

- State that active transport is the movement of a substance from an area of low concentration to an area of high concentration.

- State that active transport requires membrane proteins to move substances against the concentration gradient.

- Name sodium and potassium as examples of substances that are actively transported across nerve cell membranes.

- Name iodine as an example of a substance that is actively transported across the cell membrane in seaweed.

3 Producing new cells

You should already know:

- Cells must divide to increase the number of cells in an organism.
- Cell division is controlled by the nucleus.
- During cell division, the parent cell divides to produce two identical cells, which contain the same number of chromosomes in their nuclei as the parent cell.
- Cell division allows organisms to grow and repair damaged parts, e.g. cuts, broken bones.

Learning intentions

- Describe the stages of mitosis.
- Describe the maintenance of diploid chromosome complement by mitosis.
- Explain why mitosis is used by cells.
- Describe how cells are produced using cell-culture techniques.
- Describe the aseptic techniques that are used when culturing cells.

🌳 Biology in context

The first 22 pairs of chromosomes are called autosomes. These are common to both males and females. Pair 23 (the last pair) are the sex chromosomes. These determine your sex. Females inherit two X chromosomes, whereas males inherit an X and a Y chromosome (fig 3.2).

Fig 3.2 *The human X and Y chromosomes*

Chromosomes and cells

Within the nucleus of each body cell are 46 single chromosomes or 23 pairs of chromosomes. One chromosome from each pair is inherited from each parent. The chromosomes contain all of the genetic information.

A karyotype can be used to show the number, appearance and arrangement of chromosomes in the nucleus. The human karyotype shown in the diagram below shows how the 46 chromosomes are arranged into 23 pairs (fig 3.1).

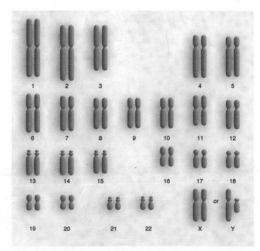

Fig 3.1 *Human karyotype*

In figure 3.3 you can see the difference between the male and female karyotypes. A cell that contains a double set of chromosomes is said to be **diploid**.

In humans a diploid cell contains 46 chromosomes—two matching sets of chromosomes, one from each parent. All body cells (except the sex cells) are diploid.

The number of chromosomes found in a cell is also known as the **chromosome complement** (one set from each parent as shown in the karyotype above). The chromosome complement varies depending on the organism.

All organisms originate from a single fertilised egg cell. Through cell division, the number of cells in an organism is increased. Cells can then become specialised to carry out specific functions within the body.

A selection of differentiated cells are shown below (figs 3.4 to 3.7):

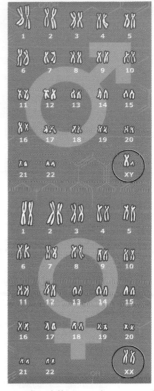

Fig 3.3 *The difference between the male and female karyotypes*

Fig 3.4 *Human sperm cell*

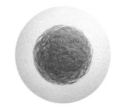

Fig 3.5 *Human egg cell*

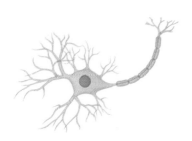

Fig 3.6 *Nerve cell*

Fig 3.7 *Red blood cell*

Make the link – Biology

You will learn more about inheritance in Unit 2, page **171**.

Make the link – Biology

You will learn more about different cell types and their functions in Unit 2, pages **114–115** and **117–118**.

GO! Activities

Activity 3.1 Working individually

(a) Write down full answers to the following questions:

1. Describe what a karyotype shows.
2. Name the term used to describe a cell with a double set of chromosomes.
3. Explain what is meant by the term 'chromosome complement'.

(*Continued*)

(b) The information in the table below shows the chromosome complement for different species.

Using the information in the table, construct a bar chart to show the chromosome complement for each of the different species.

Species	Chromosome complement	Species	Chromosome complement
Lettuce	18	Tomato	24
Cat	38	Human	46
Potato	48	Horse	64
Dog	78		

Activity 3.2 Working in pairs

Work with a partner to create a human karyotype.

- Cut out 23 chromosomes from two different-coloured sheets of paper (one colour for Mum and one for Dad).
- Make sure that each of the pairs is the same size and shape.
- Match up your 23 pairs and stick them down onto another sheet of paper.
- Label each pair 1 to 23.

You now have your own chromosome complement showing diploid pairs.

Biology in context

The majority of the human population have a chromosome complement of 46 chromosomes. However, some children are born with the genetic condition Down's syndrome.

The condition is named after the British doctor, John Langdon Down, who classified it in 1866. Down's syndrome is a result of the inheritance of an extra copy of one chromosome. This gives a chromosome complement of 47 chromosomes instead of the usual 46.

In chromosome pair 21, there are 3 chromosomes instead of 2. This extra chromosome causes physical and mental disabilities. The condition gives a characteristic appearance common to all people born with Down's syndrome (fig 3.8).

There is no cure, but with support, people with the condition can lead a relatively normal life. In the UK, around 1 out of 1000 children are born with this condition.

Fig 3.8 *A child who has Down's syndrome*

Maintaining the chromosome complement

It is important during cell division that a diploid chromosome complement is maintained. This prevents any loss of genetic information from the daughter cells. All of the genetic information needed to carry out all the cells' functions is contained within the chromosomes.

What happens to the number of chromosomes is summarised in the simple diagram of cell division shown in figure 3.9.

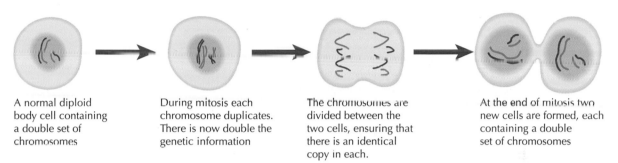

A normal diploid body cell containing a double set of chromosomes

During mitosis each chromosome duplicates. There is now double the genetic information

The chromosomes are divided between the two cells, ensuring that there is an identical copy in each.

At the end of mitosis two new cells are formed, each containing a double set of chromosomes

Fig 3.9 *Maintaining the chromosome complement*

Mitosis

The process of cell division is called **mitosis**. It takes place in animal and plant cells in order to increase the number of cells. Mitosis is required for **growth** and **repair** of cells.

It is essential that a parent cell duplicates the chromosomes contained within its nucleus before it divides. This ensures that the daughter cells produced contain the same number of chromosomes as the original cell.

Mitosis can be broken down to a number of stages as shown in fig 3.10.

 Stage 1 Mitosis starts off with a diploid parent cell. The individual chromosomes coil up and become visible.

 Stage 2 Each chromosome duplicates so that an exact copy of each chromosome is made. The chromosomes are said to consist of two chromatids joined together by a centromere.

 Stage 3 The chromosomes line up at the equator (centre) of the cell, and spindle fibres attach to each pair of chromatids.

Stage 4 The spindle fibres shorten, pulling the chromatids apart. The chromatids move towards the opposite poles of the cell.

Stage 5 The nuclear membranes re-forms around each group of chromatids and the cytoplasm divides.

Stage 6 This produces two new daughter cells, each with the same number of chromosomes as the original cell.

Fig 3.10 *The stages of mitosis*

Hint

Remember you can always use the Internet to search for short film clips of mitosis to help you understand the process

GO! Activities

Activity 3.3 Working individually

Write down full answers to the following questions:

1. Define mitosis.
2. Describe the stages of mitosis.
3. Describe how a diploid chromosome complement is maintained by mitosis.

Activity 3.4 Working in pairs

(a) Work with a partner to produce a poster showing each of the stages of mitosis. Base your poster on a cell that has a chromosome complement of 4 chromosomes.

  • Use different-coloured pens or drinking straws to represent the chromosomes.

  • Each chromosome/chromatid must be kept the same colour in each of the stages.

  • Add short descriptions to each stage to describe what is happening at each of the stages.

(b) Using six small squares of card, draw each stage of mitosis on a separate card. Mix the cards up and practise putting them into the correct order. Make sure that you can describe what happens at each of the stages. Now give them to your partner so that they can try.

Cell culture

Using the process of mitosis to grow cells in an artificial environment is called **cell culture**.

Cells will only grow inside or outside body cells under favourable conditions.

Cell production by cell culture requires:

- a suitable **growing medium** (something for them to grow on)
- availability of **oxygen**
- a suitable **temperature**
- a suitable **pH** level.

Cells can be grown in a liquid medium like **nutrient broth** in a culture vessel (this is a flat bottle that lies on its side). This type of growth medium would typically be used when culturing animal cells. Nutrient broth can also be mixed with **agar** (a type of jelly) and poured into Petri dishes. This forms a solid medium on which microbes are normally grown.

It is possible to grow layers of cells in Petri dishes or culture vessels (fig 3.11).

To provide ideal growing conditions for cells outside the body, a cell incubator is used. A sealed incubator provides a sterile environment. It also allows control over temperature, humidity, pH, carbon dioxide and oxygen levels. The cultures are placed in incubators, and under the right conditions, cells will carry out the process of mitosis.

Fermenters (fig 3.12) can also be used to grow cell cultures. They are commonly used in industry to produce cells on a large scale. Fermenters are usually made from stainless steel with nutrients and other essential materials passed in through a series of different pipelines.

Conditions inside the fermenter are controlled by computers (fig 3.13).

Fig 3.11 *Petri dishes used to grow cells*

Fig 3.12 *Fermenters*

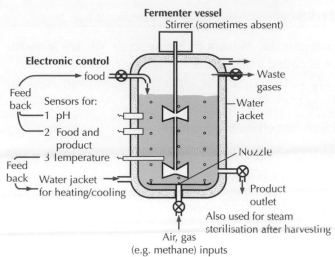

Fermenter vessel
Stirrer (sometimes absent)

Electronic control
→ food

Feed back

Sensors for:
1 pH
2 Food and product
3 Temperature

Feed back
Water jacket
for heating/cooling

→ Waste gases

Water jacket

Nozzle

Product outlet

Also used for steam sterilisation after harvesting

Air, gas (e.g. methane) inputs

Fig 3.13 A diagram of a fermenter

Fig 3.14 *A meal containing Quorn™*

Fig 3.15 *A loaf of bread*

Fig 3.16 *Antibiotics*

Fig 3.17 *Beer*

Growing bacteria and fungi in fermenters is a quick way to increase the number of cells. Making Quorn™, bread, antibiotics, beer and wine all depend upon the action of these microorganisms (figs 3.14 to 3.17).

Cells can be removed from any organisms and grown outside their natural environment using these cell-culture techniques.

A simple experiment can be carried out in the classroom to grow yeast. A small sample of liquid yeast culture is collected on a sterile inoculating loop (a thin looped wire that can be heated to a high temperature to sterilise it) and streaked across a Petri dish containing agar. The Petri dish is sealed and left in a warm place. After 24–48 hours colonies of yeast are visible (fig 3.18).

To do this experiment or any other work involving cell culture requires the use of **aseptic techniques**.

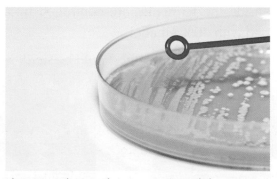

Fig 3.18 *Colonies of yeast on a Petri dish*

Aseptic techniques

Aseptic techniques involve maintaining a sterile environment. When any type of cell culture is carried out in a lab, aseptic techniques are followed. This ensures that the conditions are suitable to encourage cell division, but also prevents any contamination.

The list below gives some of the basic rules that should be followed when working with cell cultures.

Personal hygiene:

- Long hair should be tied back.

- Wear protective clothing, e.g. a lab coat, gloves and safety glasses.

- Wash your hands before and after working with cultures (fig 3.19).

Fig 3.19 *Protective clothing is essential when working with cell cultures*

Work area:

- Surfaces should be clean, tidy and well organised with only equipment out that is needed for the experiment.

- Surfaces should be disinfected before and after working with cultures.

- Ensure there are no open windows or draughts.

- A cell-culture hood (laminar-flow cabinet) should be used with proper ventilation (fig 3.20).

Fig 3.20 *A cell-culture hood*

If working with microorganisms outside a hood, a wire loop should be used and sterilised on a Bunsen burner until it is red hot (fig 3.21).

Sterile handling:

- Hands, work areas and the outside of containers should be wiped with 70% ethanol.

- All glassware and equipment should be sterilised.

- Pouring should be avoided and a separate, sterile pipette used for each liquid (fig 3.22).

- Flasks, bottles and dishes should not be left open and should only be uncovered immediately before use.

- Experiments should be performed as quickly as possible.

- All flasks, bottles and dishes should be sealed immediately after use.

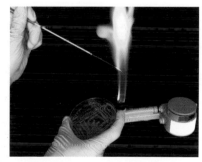

Fig 3.21 *Using a Bunsen burner to sterilise a wire loop*

Fig 3.22 *A separate, sterile pipette should be used for each liquid*

Abidance of these rules should ensure contamination is kept to a minimum.

Biology in context
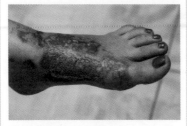

Many different types of cell have been cultured, some more successfully than others.

Human skin can be grown in a laboratory and is particularly useful for treating victims of severe burns or skin diseases. People may be at risk of dying from infection, as their skin cannot be repaired fast enough (fig 3.23).

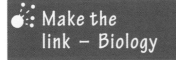

Fig 3.23 *Severe burns*

The production of artificial skin in the lab involves the use of cell-culture techniques. A small sample of skin can be removed from the patient and grown in the laboratory using cell culturing to produce sheets of skin (fig 3.24).

This produces much more skin in a shorter period of time than the body would be able to do by itself.

The sheets of skin are then grafted onto the damaged areas. This method works well, as there is less chance of rejection by the body, as the skin has come from the patient (fig 3.25).

In cases where it is not possible to remove healthy skin from patients, alternative methods can be used. Stem cells taken from the umbilical cord of newborn babies and, more recently, skin from the foreskins of circumcised baby boys have been used to grow artificial skin in a laboratory. Do an internet search to find videos showing how new skin can be grown in a laboratory.

Make the link – Biology

You can find out more about stem cells in Unit 2 (page **120**)

Fig 3.24 *A sheet of artificial skin*

Fig 3.25 *Performing a skin graft*

GO! Activities

Activity 3.5 Working individually

Write down full answers to the following questions:

1. Explain what is meant by the term 'cell culture'.
2. When any type of cell culture is carried out in a laboratory, aseptic techniques are followed. Explain what is meant by the term 'aseptic techniques'.
3. Name some of the factors that must be controlled during the growth of cell cultures.

Activity 3.6 Working individually

The table below gives the average time taken for one cell to complete cell division.

Cell type	Time taken
Fly embryo	8 minutes
Bacteria	20 minutes
Yeast	2 hours
Human skin	24 hours

Use the table to complete the tasks below.

1. Copy and complete the table below to show how many cells are present in a fly embryo after a period of 40 minutes.

Time (minutes)	Number of cells
0	1
8	

2. Calculate how long it will take to produce 128 yeast cells.

3. Construct a table to show how many bacterial cells would be present after a period of 3 hours.

4. Using your own table of results, produce a line graph to show the number of bacterial cells produced over a 3-hour period.

Activity 3.7 Working individually

Carry out some research of your own on growing cells by cell culture. Produce an information leaflet detailing the steps involved. Make sure that you include any safety procedures that must be followed.

Activity 3.8 Working individually

Current research involves whole organs being grown in a laboratory. See what you can find out about this from Internet research. Produce a 200-word report summarising your findings.

I can:

- State that diploid cells are cells that have two matching sets of chromosomes.

- State that each human body cell is diploid (contains 46 chromosomes).

- State that the process of mitosis maintains the diploid chromosome complement.

- Describe the importance of maintaining a diploid chromosome complement.

- Describe each of the stages of mitosis (including the movement of chromatids to the equator of the cell and the formation of spindle fibres).

- State that mitosis is required for growth and repair.

- State that cells can be grown outside of their natural environment using cell-culture techniques.

- State that the culturing of cells requires the use of aseptic techniques.

- Describe aseptic techniques that are used when culturing cells.

- Give examples of growth media used to grow cells, e.g. nutrient broth and agar.

- State that factors such as oxygen, temperature and pH are controlled during the growth of cell cultures.

4 DNA and the production of proteins

You should already know:

- Genes are located on chromosomes in the nucleus.
- Genes are made of DNA.
- DNA carries instructions to make proteins.
- Each individual's DNA is unique.
- Genes are passed on from parents to offspring.

Learning intentions

- Describe the structure of DNA.
- State the names of the four bases that make up the genetic code.
- Explain the relationship between DNA and proteins.
- Explain the relationship between the order of bases on DNA and the amino acids in a protein.
- Describe the role of mRNA in protein production.
- Name the basic units that proteins are made from and where protein synthesis takes place.

Chromosomes, genes and DNA

All of the genetic information is held on chromosomes found within the nucleus of almost all body cells. The chromosomes are made up of a chemical called deoxyribonucleic acid (DNA).

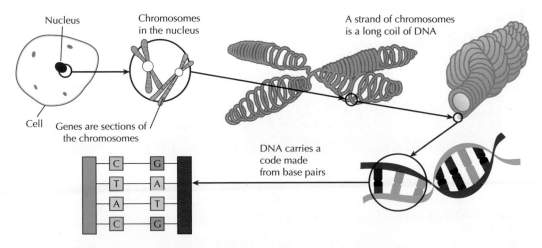

Nucleus

Chromosomes in the nucleus

A strand of chromosomes is a long coil of DNA

Cell

Genes are sections of the chromosomes

DNA carries a code made from base pairs

Fig 4.1 *How DNA is packaged into a cell. Where a black circle is shown, this indicates an area that has been magnified*

If we look at a chromosome in greater detail we can see the bands of different genes. The genes are made up of DNA, which we will examine in greater detail in this chapter.

The structure of DNA

DNA is a molecule that consists of two strands connected together by bases. The DNA is described as a **double-stranded helix,** as it is found twisted and tightly packed together, rather like a coiled spring (fig 4.2).

The two strands are made up of chemicals called **bases**. There are four different bases found in DNA; adenine (**A**), thymine (**T**), guanine (**G**) and cytosine (**C**). The two strands of DNA join together when the bases join together. The bases always pair together in the same way; A joins with T, and G with C. This is known as **complementary base pairing** (fig 4.3). The bases make up the **genetic code**.

🌳 The History of DNA

Throughout history, scientists have been making discoveries about DNA. It was first isolated from the cell nucleus in 1869 by a biologist called Friedrich Miescher. Research continued by various scientists, and by the late 1940s it was accepted that DNA was the molecule of life. However, the structure of the molecule was then unknown.

It wasn't until 1953 that molecular biologists James Watson and Francis Crick produced a model showing the structure of DNA as a double helix (fig 4.4).

Fig 4.2 *Model of DNA*

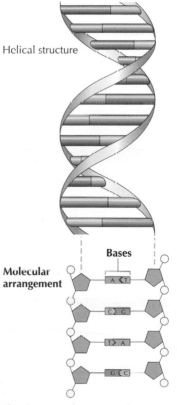

Helical structure

Molecular arrangement

Bases

Fig 4.3 *Complementary base pairing in DNA*

Fig 4.4 *Watson and Crick with a model showing the structure of DNA as a double helix*

(*Continued*)

They were later awarded a Nobel prize for the discovery, along with Maurice Wilkins. Watson and Crick based their ideas on earlier research carried out by Rosalind Franklin and Wilkins. Both Franklin and Wilkins carried out a lot of work on X-ray diffraction of DNA. It was an X-ray diffraction photograph of Franklin's that helped Watson and Crick to understand the helical structure of the molecule.

The base-pairing rule had first been explained in 1949 by a biochemist named Erwin Chargaff. He discovered that although different organisms had different quantities of DNA, the percentage of adenine was always equal to the percentage of thymine, as was the percentage of cytosine equal to the percentage of guanine.

Extracting DNA

DNA can be extracted from fruit and vegetables. The experiment below describes how DNA was extracted from a strawberry.

Method

1. A strawberry is added to a plastic bag and smashed to a pulp.
2. An extraction fluid is made by combining the following: 90 ml of water, 5 g of salt and 10 ml of detergent.
3. 30 ml of extraction fluid is added to the bag containing the strawberry and mixed well.
4. The bag is placed in a 60°C water bath for 15 minutes.
5. The bag is then placed into an ice bath for 5 minutes.
6. The mixture is filtered to remove any remaining solid strawberry.
7. The mixture is poured into a boiling tube until it is half full.
8. The boiling tube is tilted to the side, and ice-cold ethanol is slowly poured down the side of the tube.

Fig 4.5 *DNA extracted from a strawberry*

Result

The DNA is visible floating on top of the ethanol (fig 4.5).

⊙ Activities

Activity 4.1 Working in pairs

Firstly, use a mini whiteboard to 'thought-shower' what you know about DNA. Make a simple diagram showing DNA as a double-stranded molecule with four different bases.

Using this information, choose from **one** of the following tasks:

• Create a DNA model.

Use any material available to you. Suggested materials are: coloured plasticine/ Play-Doh, coloured pipe cleaners/drinking straws, coloured paper or sweeties that could be linked together with cocktail sticks. The choice is yours!

• Produce a colourful DNA poster.

Whichever task you choose, you must show the DNA as a double-stranded molecule and the four different bases. You must also show complementary base pairing. You do not have to show the DNA as a twisted double helix, but you must show that it is double-stranded (fig 4.6).

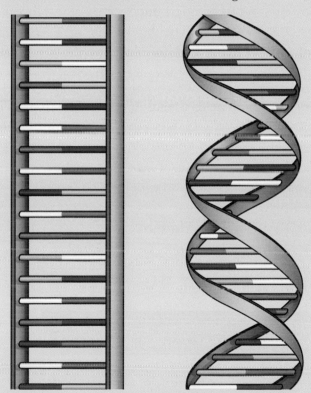

Fig 4.6 *DNA untwisted and twisted into a double helix*

Activity 4.2 Working individually

Carry out some research into the history of DNA. Produce a timeline of the discovery of the molecule and a short biography of the main scientists involved.

DNA and proteins

The DNA contains the genetic information for making **proteins**. Proteins are made up of amino acid molecules joined together. A group of three bases on the DNA form a triplet.

Each triplet creates a code for a different amino acid (fig 4.7).

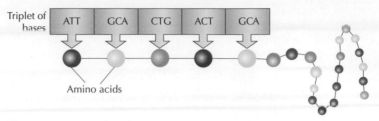

Triplet of bases | ATT | GCA | CTG | ACT | GCA

Amino acids

Fig 4.7 *Amino acid triplets*

As there are four different bases (A, T, G and C), the triplet code allows for there to be a different code for each of the 20-plus different amino acids that make up all the proteins in nature.

Each amino acid is represented by three letters – the abbreviated name of the amino acid, e.g. 'Gly' is short for 'glycine'.

The **type of protein made is determined by the sequence of bases** on the DNA. The sequence of bases on the DNA will provide the information for which amino acids are coded for. The **order of the amino acids then determines which protein is formed**. As each individual's DNA is different, the proteins coded for are unique to that individual.

Make the link – Biology

All humans have physical differences. We can account for these differences owing to our DNA coding for the different proteins made in our bodies. You will examine some of these differences in Unit 2 when you study variation and inheritance (Chapter 13, page **156**).

Hint

When drawing bar charts remember the following:

- labels on both axes (including units)
- scales should usually start at zero
- bars should be of equal width and equally spaced
- always use a ruler!

GO! Activities

Activity 4.3 Working individually

(a) A section of DNA has 800 bases. Copy and complete the table below to show how many of each type of base there are and the percentage of each type of base.

Type of base	Number of bases	Percentage of bases
A	160	
T		
G	240	
C		30

(b) Using the information in the table above, draw a bar chart to show the percentage of each type of base.

Production of proteins

The information to produce proteins is held on the DNA in the nucleus. Proteins carry out a variety of jobs in the body and are essential for life. They are produced at ribosomes found in the cytoplasm of the cell.

As there is no way for the DNA to leave the nucleus, a copy of the code must be taken from the DNA in the nucleus.

Messenger RNA (**mRNA**) is a molecule that **carries a copy of the code from the DNA** held in the **nucleus** to a **ribosome** in the cytoplasm. From the copied code, the correct **amino acids are joined together**. The correct amino acid to match the code is identified using the complementary base-pairing rule. The amino acids form a long chain and eventually **a protein is formed** (fig 4.8).

The variety and functions of proteins are discussed in the next chapter.

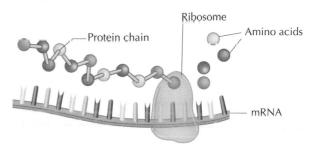

Fig 4.8 *Protein chain being formed at a ribosome from mRNA code*

🔵 Activities

Activity 4.4 Working individually

Write down full answers to the following questions:

1. Describe the structure of DNA.
2. Name the four bases found in DNA.
3. Explain what the base sequence in DNA determines.
4. State the function of the molecule mRNA.

Activity 4.5 Working individually

Make a mini poster to show a protein chain.

- From the 4 bases (A, T, G and C), make up a section of DNA consisting of 15 bases in any order of your choice. Show this on your poster.
- Use a different colour for each amino acid and show how a protein chain is formed.

I can:

- Describe the structure of DNA using the terms: double helix; bases.

- Identify that DNA has 4 different bases: Adenine, Thymine, Guanine and Cytosine.

- Identify that base A pairs with base T, and base G pairs with base C; this is known as complementary base pairing.

- State that DNA carries the genetic information for making proteins which are unique to each individual.

- Explain that the sequence of bases determines the sequence of amino acids in a protein.

- State that messenger RNA (mRNA) is a molecule that carries a copy of the code from the DNA in the nucleus to a ribosome in the cytoplasm.

- State that at a ribosome, the mRNA is decoded, and the correct amino acids are assembled from the code.

- State that amino acids join together to form a protein.

5 Proteins and enzymes

You should already know:

- Enzymes are found in living cells.
- They speed up chemical reactions in cells, are specific and remain unchanged following a chemical reaction.
- Enzymes build up and break down molecules.
- Enzymes can be used in a range of biotechnology industries.

Learning intentions

- Explain how the variety of protein shapes and functions arises.
- Describe some of the main functions of proteins.
- State what enzymes are and where they can be found.
- Describe the main function of an enzyme.
- Define the terms 'active site' and 'substrate' in relation to enzymes.
- Explain the relationship between the active site of an enzyme and its substrate.
- Explain the meaning of the term 'optimum' as applied to enzymes.
- Give the factors that affect enzymes and other proteins, and describe their effect.
- Explain the meaning of the term 'denatured' and why it happens to enzymes.

Variety and function of proteins

Fig 5.1 *Protein as a 3-dimensional shape*

Proteins are complex molecules that contain the chemical elements carbon, hydrogen, oxygen and nitrogen. There are many different types of proteins, and when they are examined at a molecular level they all have very different 3-dimensional shapes. Special computer programs are used to show the 3-dimensional shapes of proteins. An example of this is shown in figure 1.67. The protein looks very ribbon-like.

The variety of proteins is due to the **sequence of amino acids** which is determined by the order of bases on the DNA. As there are 20 different amino acids, there many different ways in which they can be arranged.

Amino acids are joined together in a chain connected by peptide bonds. Long chains of amino acids form a polypeptide chain. The polypeptide chains twist around, forming many different shapes. Different combinations of amino acids can join together, making the possible combinations of proteins almost endless.

To illustrate this in a simple way, different shapes are often used to represent the different amino acids. Figure 5.2 shows how different amino acids are joined together to form proteins.

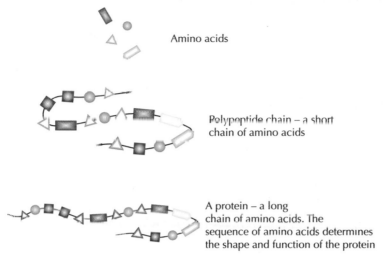

Amino acids

Polypeptide chain – a short chain of amino acids

A protein – a long chain of amino acids. The sequence of amino acids determines the shape and function of the protein

Fig 5.2 *Amino acids joined together to form a protein*

Proteins have many functions in the human body. One of these functions was explored in pages 24–25 of chapter 2 when the structure of the cell membrane was examined. Proteins were shown to have an important **structural** function in cell membranes. Proteins are needed to support the membrane. Some have channels, and others act as carriers to allow the movement of substances into and out of cells (fig 5.3).

Some of the membrane proteins act as **receptors**. Receptors receive external signals and provide a binding site for molecules. This allows information to be passed to the inside of the cell.

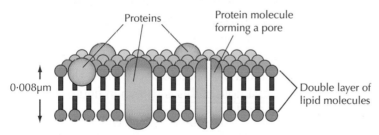

Proteins

Protein molecule forming a pore

0·008µm

Double layer of lipid molecules

Fig 5.3 *Structural function of proteins in the cell membrane*

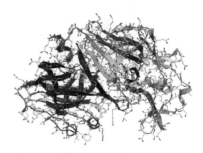

Fig 5.4 *3-dimensional structure of the enzyme pepsin*

Other proteins act as **enzymes**. These are important in cell metabolism – all of the chemical reactions that occur in cells. Enzymes are necessary to speed up the chemical reactions that are essential for life. The enzyme pepsin is important to break down proteins in the body. Its structure is shown in fig 5.4.

Proteins also function as **hormones**. Hormones are chemical messengers that are important in the regulation of processes like growth in flowering plants, and puberty in humans. The structure

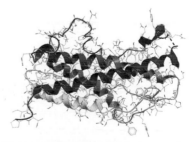

Fig 5.5 *3 dimensional structure of human growth hormone*

of human growth hormone is shown in figure 5.5. You can see how the structures of all proteins differ.

The body produces **antibodies**, which are made of protein to fight against disease. Antibodies are produced in response to specific disease-causing microorganisms. A different antibody molecule is required for each microorganism. The structure of immunoglobulin (a type of antibody) is shown in figure 5.6.

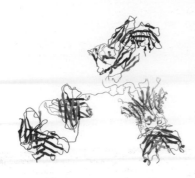

Fig 5.6 *3-dimensional structure of immunoglobulin*

> ### GO! Activity
>
> **Activity 5.1 Working individually**
>
> **(a)** Write down full answers to the following questions:
>
> 1. Explain how the variety of protein shapes and functions arises.
>
> 2. Describe the five main functions of proteins.
>
> **(b)** Draw 4 different protein chains – one for each of the protein functions described above. Each chain should have at least 6 amino acids in it. Use a single line to join the amino acids to each other to represent the peptide bonds between them. Label your diagram as fully as possible.

Enzymes and their functions

Enzymes are biological catalysts made by all living cells. A catalyst is a substance that speeds up a chemical reaction. An enzyme is classed as a **biological catalyst**, as it speeds up reactions that take place inside cells.

Although enzymes **speed up cellular reactions**, they themselves remain **unchanged** by the process and can be used over and over again.

Enzyme specificity

Each enzyme can catalyse only one reaction. This is due to their specific nature. Enzymes are said to be specific, as they can only act on one substance, their **substrate**.

The unique shape of each enzyme means that it can join with only one substrate. This is often called the 'lock and key' hypothesis. On each enzyme there is an active site (the lock). This is the place where the chemical reaction takes place. Only one substrate (the key) can fit together with the active site. The shape of the **active site** on the enzyme is **complementary** to the shape of the **substrate**.

When a substrate binds to an enzyme's active site, it forms an 'enzyme – substrate complex'. Only when this complex is formed will the reaction take place. The reaction can involve the build-up or breaking down of substances. Once the reaction has taken place, the product(s) of the reaction are released, and the enzyme moves on to carry out the same reaction with more of the same substrate. This continues until there is no more of the substrate left (fig 5.7).

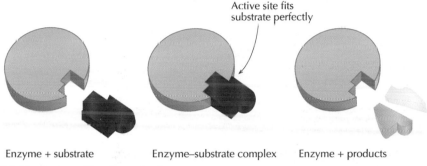

Active site fits substrate perfectly

Enzyme + substrate   Enzyme–substrate complex   Enzyme + products

Fig 5.7 *The lock and key hypothesis*

Investigating enzyme action

An example of an enzyme found in all living cells is catalase. It speeds up the breakdown of hydrogen peroxide. Hydrogen peroxide is a poisonous substance, and its build-up inside the body would result in death.

Hydrogen peroxide can be broken down into oxygen and water – two harmless substances. To ensure that the reaction works fast enough to prevent the body from coming to harm, the enzyme catalase makes the reaction go faster.

The breakdown of hydrogen peroxide is demonstrated by adding a piece of liver tissue to some hydrogen peroxide. Bubbles of oxygen gas are produced (fig 5.8).

Fig 5.8 *The breakdown of hydrogen peroxide by the enzyme catalase*

The reaction is summarised in the following word equation:

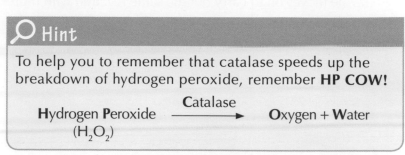

$$\text{Hydrogen peroxide} \xrightarrow{\text{Catalase}} \text{Oxygen + water}$$
$$(H_2O_2)$$

Substrate   Enzyme   Products

🔍 Hint

To help you to remember that catalase speeds up the breakdown of hydrogen peroxide, remember **HP COW!**

$$\textbf{H}\text{ydrogen } \textbf{P}\text{eroxide} \xrightarrow{\text{Catalase}} \textbf{O}\text{xygen + } \textbf{W}\text{ater}$$
$$(H_2O_2)$$

GO! Activities

Activity 5.2 Working individually

Write down full answers to the following questions:

1. State what enzymes are and where they can be found.
2. Describe the main function of an enzyme.
3. Define the terms 'active site' and 'substrate' in relation to enzymes.
4. Explain the relationship between the active site of an enzyme and its substrate.

Activity 5.3 Working in pairs

Read the information below carefully and discuss the experiment.

A pupil carried out an experiment to find out which type of tissue contained the most catalase.

The experiment was set up as follows:

- 10 ml of hydrogen peroxide and 2 drops of detergent were added to 4 boiling tubes as shown below.
- Three discs of a different type of tissue were added to each boiling tube: carrot, apple, potato and boiled potato (fig 5.9). The shape, age and mass of all the tissues were kept the same.

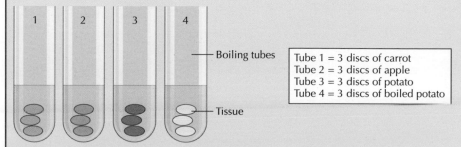

Boiling tubes

Tube 1 = 3 discs of carrot
Tube 2 = 3 discs of apple
Tube 3 = 3 discs of potato
Tube 4 = 3 discs of boiled potato

Tissue

Fig 5.9 *Investigating catalase concentration in different types of tissue*

- The boiling tubes were left for 5 minutes, and then the height of foam produced in each tube was measured.
- The results are shown in the table below:

Type of tissue	Height of foam (mm)
Carrot	58
Apple	42
Potato	64
Boiled potato	0

1. Using the results in the table above, draw a bar chart to show the height of foam for each of the tissue types.
2. Read the information below and discuss it with your partner.

In any experiment a scientist must ensure the following:

- **Validity**: This is to do with the 'fairness' of the experimental procedure. In any experiment, only one factor (variable) should be varied, and all other factors (variables) that could affect the results should be kept the same.
- **Reliability**: To ensure reliability of an experiment and the results, the experiment is repeated, and several readings or samples are taken.
- **Controls**: A control is identical to the original experiment in every way except that the factor that causes the change is removed. This is then replaced by something that does not bring about change.

Write down full answers to the following questions:

(i) What did the pupil do to ensure the validity of the experiment?

(ii) Which variables were kept the same?

(iii) What did/could the pupil do to ensure the reliability of the procedure?

(iv) Was a control used in the experiment? What would it tell us?

(v) What conclusions can be drawn from this experiment?

Activity 5.4 Working individually

Read the information below carefully, and then complete the tasks that follow.

A pupil carried out an experiment to investigate the specificity of enzymes.

The experiment was set up as follows:

- 10 ml of hydrogen peroxide and 2 drops of detergent were added to 4 boiling tubes (fig 5.10). The same bottle of hydrogen peroxide and detergent was used each time, and all boiling tubes were of the same diameter.
- 1 ml of a different enzyme was added to each test tube: amylase (tube 1), catalase (tube 2), lipase (tube 3), trypsin (tube 4).

(Continued)

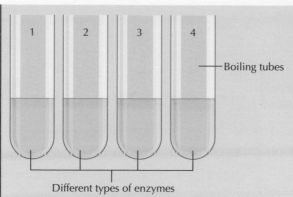

Boiling tubes

Different types of enzymes

Fig 5.10 *Investigating enzyme specificity*

- The boiling tubes were left for 5 minutes, and then the height of foam produced in each tube was measured.
- The results are shown in the table below:

Enzyme	Height of foam (mm)
Amylase	0
Catalase	50
Lipase	0
Trypsin	0

Write down full answers to the following questions:

(a) What conclusions can be drawn from the results obtained?

(b) What has been done to ensure the validity of the experiment?

(c) Which variables were kept the same?

(d) What could be/is done to ensure the reliability of the procedure?

(e) Was a control used in the experiment? If not, what would a suitable control be? What would it tell us?

Enzyme action

Each enzyme works best under **optimum** conditions. At its optimum an enzyme is at its most active – the rate of reaction cannot go any faster. Although the optimum conditions vary for each enzyme, all enzymes are affected by **temperature** and **pH**. These factors affect enzymes and other proteins by causing changes to their shape.

If the shape of an enzyme's active site is altered it is said to be **denatured** and will no longer fit together with its substrate.

If an enzyme-substrate complex cannot be formed, this will affect the rate of the reaction. Without an enzyme to speed up the reaction it will take place at a much slower rate.

The experiment shown in figure 5.11 illustrates the effect of different temperatures on the action of catalase.

Liver tissue (source of catalase) is added to hydrogen peroxide that ranges in temperature from 5°C to 60°C. Catalase speeds up the breakdown of hydrogen peroxide, releasing oxygen bubbles.

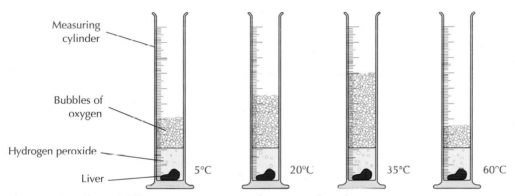

Measuring cylinder

Bubbles of oxygen

Hydrogen peroxide

Liver

5°C   20°C   35°C   60°C

Fig 5.11 *The effect of different temperatures on the action of catalase*

At 5°C and 60°C there is decreased oxygen production, as the conditions are not optimal for catalase activity. At 35°C most oxygen bubbles are produced.

The optimum temperature and pH of each enzyme are normally linked to the organism in which it is found. Most enzymes found in the human body work best at 37°C (body temperature). Plant enzymes on the other hand work best around 20°C.

At low temperatures, the rate of a reaction is decreased as enzyme molecules have less energy to move around. As the temperature is increased, the rate of enzyme activity increases until it reaches the optimum (fig 5.12).

After the optimum, as temperature continues to increase the rate of enzyme activity decreases. The increased temperature causes the shape of the active site to change. This makes it difficult for the substrate to join with the enzyme. If the temperature continues to increase (over 50°C), the shape of the active site is permanently changed and the enzyme is **denatured**.

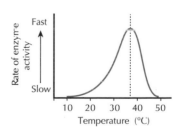

Fig 5.12 *Enzyme activity and temperature*

The optimum pH of an enzyme can vary depending upon where it functions in the body. The enzyme pepsin works best around pH 2. It is found in the stomach where conditions are very acidic. Catalase found in the liver works best at pH 9 (fig 5.13).

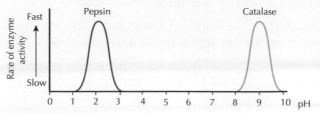

Fig 5.13 *Enzyme activity and pH*

Graphs of enzyme activity with increasing temperature or pH all have a similar shape. Enzyme activity increases until the optimum and then decreases (fig 5.14).

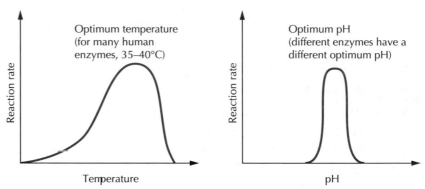

Fig 5.14 *Graphs of optimum temperature and pH*

Biology in context

You may or may not have been aware of it, but at some point you will most likely have observed the effects of temperature on proteins.

If you have boiled, fried or poached an egg, you will have observed the effects of temperature on a protein.

The white of an egg (egg albumen) is made of protein. In a raw egg (fig 5.15a) the albumen is a clear, runny liquid. As the albumen is heated, its consistency changes to become a white solid (fig 5.15b). Once heated, the albumen is permanently altered and cannot be changed back.

Fig 5.15a *Raw egg*

Fig 5.15b *Fried egg*

Enzymes are also made of protein and so are irreversibly altered once heated. When heated to over 50°C the shape of the active site is permanently changed. This causes the enzyme to become denatured.

It is important to remember that although enzymes are made of protein, not all proteins are enzymes.

GO! Activities

Activity 5.5 Working individually

Read the information below carefully, and then complete the tasks that follow.

An experiment was carried out to investigate the effect of pH on enzyme activity.

Hydrogen peroxide is used as the substrate and catalase as the enzyme. Catalase speeds up the breakdown of hydrogen peroxide and is most active at its optimum pH. The breakdown of hydrogen peroxide results in the formation of oxygen, so the height of foam produced is used as a measure of enzyme activity.

The experiment was set up as follows:

- 10 ml of hydrogen peroxide and two drops of detergent were added to six boiling tubes.
- 2 ml of six different pH solutions were mixed with the hydrogen peroxide ranging from pH 6 to 11.
- A 2 gram piece of liver was then added to each boiling tube (fig 5.16).

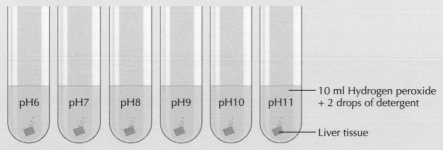

Fig 5.16 *Effect of pH on the breakdown of hydrogen peroxide*

(Continued)

- The boiling tubes were left for 5 minutes, and then the height of foam produced in each tube was measured.
- The results are shown in the table below.

pH	Height of foam (mm)
6	0
7	15
8	30
9	20
10	15
11	0

Complete the following tasks:

1. Using the information given in the table above, draw a line graph to show the height of foam produced at each pH.
2. Using the information given in the table state the pH at which the enzyme reaches its optimum.

Activity 5.6 Working individually

Produce an information leaflet about enzymes. You must include the following information:

- what enzymes are and where they are found
- the main function of an enzyme
- define the terms 'active site' and 'substrate'
- explain the relationship between the active site of an enzyme and its substrate.
- a diagram to illustrate the 'lock and key' hypothesis
- give the meaning of the term 'optimum'
- name the factors that affect enzymes, and describe their effect
- define the term 'denaturation', and explain when it happens

Activity 5.7 Working in pairs

Carry out some research into enzymes and their optimum conditions. Find out about as many enzymes as you can. Construct a table of the information listing the name of the enzyme, its substrate, optimum temperature and optimum pH.

🌳 Biology in context

Enzymes have many industrial uses.

Many products and processes we use on a daily basis depend upon the action of enzymes. The production of cheese, baking of bread, brewing of beer and wine, and breakdown of waste are just some examples.

Enzymes are also used in biological washing powders to digest protein, starch and fat stains. Wash cycles normally use temperatures of 40–90°C. As enzymes work well at lower temperatures clothes can be washed on cooler wash cycles (30–40°C) and be cleaned just as well as or better than with non-biological washing powders. This is better for the environment and helps to save money.

Fig 5.17 *Enzymes are used in biological washing powders*

🔍 Hint

You may wish to use information you have gathered from researching enzymes to complete a short scientific report (50–100 words) to generate evidence for Assessment Standards 2.2 and 2.3.

Use enzymes as an example of an application of biology to produce a report. This should describe the use of the application and consider the impact of this application on the environment/society.

⚛ Make the link – Biology

In Unit 2 you will learn about the digestive system. Enzymes play an important role in the digestion of food.

I can:

- Explain that the variety of protein shapes and functions arises from the sequence of amino acids.
- Describe the functions of different proteins as structural, receptors, enzymes, hormones and antibodies.
- State that enzymes are biological catalysts and are made by living cells.
- State that enzymes speed up cellular reactions but remain unchanged by the process.
- State that the active site is the site of the chemical reaction on an enzyme.
- State that a substrate is the substance that is acted upon by an enzyme.
- Explain that the substrate binds to an enzyme at its active site, and the shape of the active site is complementary to only one specific substrate.
- Explain that optimum conditions are the conditions at which enzymes are most active.
- State that factors such as temperature and pH affect enzymes and other proteins, and result in changes to their shape.
- Explain that at high temperatures enzymes are denatured, resulting in a change to the enzyme's active site, which affects the rate of a reaction.

6 Genetic engineering

Genetic transfer

Genetic information can be **transferred** from one cell to another **naturally**.

In humans, genetic information is replicated naturally during the process of mitosis (Chapter 3, page 38). Bacterial cells contain DNA as well as circular plasmids (Chapter 1, page 14). When a bacterial cell is replicated, the DNA and plasmids inside the cell are also replicated. A **bacterial plasmid** is a small DNA molecule that can be replicated independently. Often bacterial plasmids contain genetic information that is of survival value to the organism, such as antibiotic resistance. Plasmids can be transferred from one bacterial cell to another naturally. This passes on the useful genetic information (fig 6.1).

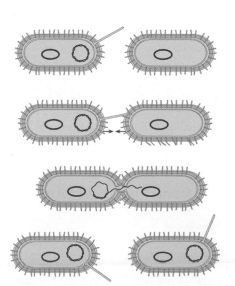

Fig 6.1 *Transfer of plasmid from one bacterium to another*

Living host cells are required for **viruses** to replicate. When a virus attacks a cell it injects its genetic information into the host cell. The genetic information from the virus can then be incorporated into the host cell's DNA. In this way there has been a transfer of genetic information. Whenever the host cell replicates, the viral genetic information will also be replicated. This also happens naturally.

Scientists have used this knowledge to artificially alter the genetic code of organisms through a process called genetic engineering or genetic modification (GM).

> **Make the link – Biology**
>
> You will find out more about the impact of genetic engineering when GM crops are investigated in Unit 3, pages **316–317**.

Genetic engineering

Genetic engineering involves the transfer of genes from one organism to another. Scientists have used this process to genetically modify food and for the production of medicines.

Some examples include (figs 6.2 – 6.7):

Fig 6.2 *Golden rice*

Fig 6.3 *Less toxic rapeseed oil*

Fig 6.4 *Bird resistance to bird flu*

Fig 6.5 *Tomatoes with a longer shelf-life*

Fig 6.6 *Blight-resistant potatoes*

Fig 6.7 *Production of medicines, e.g. insulin and human growth hormone*

Stages in genetic engineering

The steps involved in the process of genetic engineering are shown below.

1. Identify the section of DNA that contains the required gene from the source chromosome.
2. Extract the required gene.

Fig 6.8 *Diabetic injecting the hormone insulin*

3. Insert the required gene into a vector/bacterial plasmid.

4. Insert the plasmid into a host cell.

5. Grow the transformed cell to produce the required product/protein.

The process of genetic engineering is used to produce insulin for humans. Insulin is normally made in the pancreas. It is a hormone (made of protein) that controls the levels of glucose (sugar) in the blood. Some people are unable to make insulin; this is called type 1 diabetes. Diabetics need to inject insulin to control their blood sugar level (fig 6.8).

Before genetic engineering, insulin was extracted from pigs. Not only was it a slow and expensive process, but pig insulin is not exactly the same as human insulin. Some people suffered from side effects, and others were uncomfortable with the use of animals.

Figure 6.9 shows the stages involved in the production of insulin using genetic engineering.

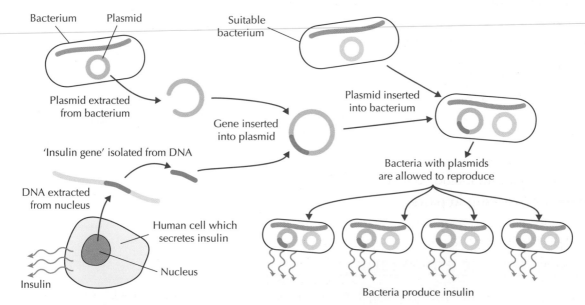

Fig 6.9 *Stages of genetic engineering*

- The chromosome containing the insulin gene is extracted from a human cell, and the insulin gene is removed using enzymes.

- A plasmid is extracted from a bacterial cell and cut open using enzymes.

- The insulin gene in inserted into the plasmid using enzymes. The plasmid now contains the genetic information to make insulin, as well as the normal bacterial information.

- The plasmid is inserted back into the bacterial cell, which is allowed to reproduce.

- The genetically modified bacteria are grown in industrial fermenters to produce large volumes of insulin.

- The insulin can then be extracted and purified from the bacterial cells and used to treat diabetics.

Make the link – RMPS

The technology used in genetic engineering raises a number of ethical issues.

Make the link Geography

The applications of genetic engineering in agriculture, food production and raw materials have an impact on economic activity, the environment and sustainability.

GO! Activities

Activity 6.1 Working individually

Construct a table to show the list of genetically engineered food and medicines mentioned in the text. For each example explain the benefit to humans of the genetic modification.

Activity 6.2 Working individually

Investigate the arguments for and against genetic engineering. Produce an essay that clearly states both points of view. After careful consideration of both sides of the argument state your own personal viewpoint on the topic.

Activity 6.3 Working in pairs

Produce a poster to show the stages involved in genetically altering an organism. Include diagrams.

Activity 6.4 Working in pairs

Use the Internet to research some of the more extreme examples of genetic engineering, e.g. Enviropig, venomous cabbage, glow-in-the-dark cats and medicinal eggs.

Produce a fact file giving 4 examples of extreme genetic engineering.

Include the following information:
- The name of the organism
- An explanation of how the organism has been genetically modified
- What the benefits are of the modification.

Hint

Plasmids are also called vectors when they are used in genetic engineering.

Hint

You may wish to use information you have gathered from researching genetic engineering to complete a short scientific report (50–100 words) to generate evidence for Assessment Standards 2.2 and 2.3.

I can:

- State that genetic information can be transferred from one cell to another naturally (by bacterial plasmids or viruses) or by genetic engineering.

- Explain the process of genetic engineering by giving details of the stages involved: identifying section of DNA that contains the required gene from the source chromosome; extracting the required gene; inserting the required gene into the vector/bacterial plasmid; inserting the plasmid into the host cell and growing the transformed cells to produce a GM organism.

7 Photosynthesis

The process of photosynthesis

Photosynthesis is the process carried out by all green plants to make food. The word photosynthesis comes from the Greek language for photo, meaning 'light', and synthesis, meaning 'putting together'.

The raw materials – carbon dioxide and water – are used to produce sugar (in the form of glucose) and oxygen. Also essential for the process to take place is light energy (from the Sun) and chlorophyll (the green pigment found in chloroplasts).

Figure 7.1 gives an overview of photosynthesis and the summary word equation.

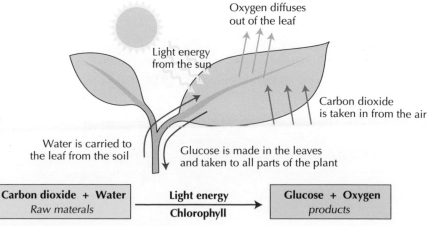

Oxygen diffuses out of the leaf

Light energy from the sun

Carbon dioxide is taken in from the air

Water is carried to the leaf from the soil

Glucose is made in the leaves and taken to all parts of the plant

Carbon dioxide + Water	Light energy	Glucose + Oxygen
Raw materals	Chlorophyll	*products*

Fig 7.1 *Overview of photosynthesis*

Photosynthesis experiments

The sugar produced by plants during photosynthesis is converted to starch.

The following procedure is used to prove that plants store sugar as starch.

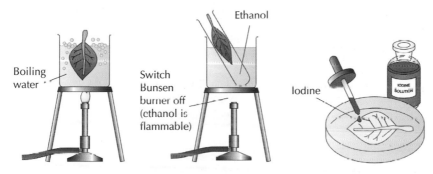

Fig 7.2 *Procedure used to test leaves for starch*

1. A leaf is removed from a plant and boiled in water for a few minutes to soften it.

2. The leaf is then placed in a test tube of ethanol, which is placed into the hot water. The ethanol removes the chlorophyll from the leaf.

3. The leaf is rinsed in water.

4. The leaf is tested for the presence of starch using iodine solution.

5. Iodine solution changes colour from red-brown to blue-black in the presence of starch. See figure 7.3.

Fig 7.3 *A leaf before and after testing with iodine*

The following experiments demonstrate the conditions required by a plant for photosynthesis to take place and the production of oxygen during photosynthesis.

Light is required for photosynthesis

A plant is placed in the dark for 24 hours so that all stores of starch are used up.

Part of a leaf is covered with aluminium foil, and the plant is left in sunlight.

After a few hours the leaf is tested for starch. Only the areas exposed to sunlight are found to contain starch.

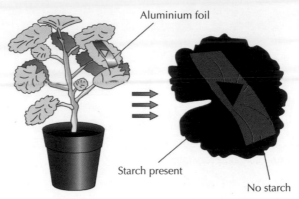

Fig 7.4 *Experiment to show the effect of light on photosynthesis*

Carbon dioxide is required for photosynthesis

A plant is placed in the dark for 24 hours so that all stores of starch are used up.

The plant is then enclosed in a plastic bag that contains the chemical sodium hydroxide. The sodium hydroxide absorbs the carbon dioxide (fig 7.5). After a few hours the leaf is tested, and no starch is found to be present.

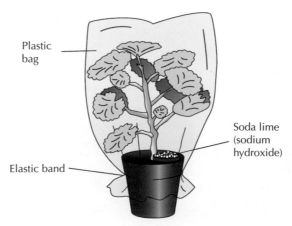

Fig 7.5 *Experiment to show the effect of carbon dioxide on photosynthesis*

Chlorophyll is required for photosynthesis

A plant with variegated leaves (fig 7.6) is placed in the dark for 24 hours to remove the stored starch. The plant is then placed in sunlight for a few hours. When the leaf is tested for starch, only the green parts of the leaf are found to contain starch (fig 7.7).

Fig 7.6 *Variegated leaf*

Before   After

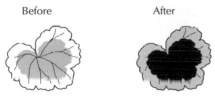

Fig 7.7 *Starch test on variegated leaf*

Oxygen production during photosynthesis

Oxygen production can also be observed in aquatic plants such as Canadian pondweed (Elodea). If the plant is kept in well-lit conditions, bubbles of oxygen are seen to be released from the stem of the plant. The oxygen produced can be collected and the gas tested (fig 7.8).

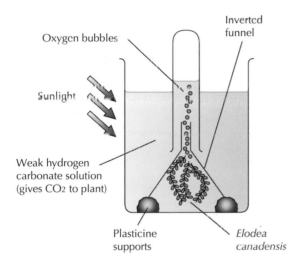

Oxygen bubbles

Inverted funnel

Sunlight

Weak hydrogen carbonate solution (gives CO_2 to plant)

Plasticine supports

Elodea canadensis

Fig 7.8 *Oxygen production by Elodea canadensis*

Photosynthesis is often thought of as a single reaction, but it is in fact a series of **enzyme-controlled** reactions that take place in a **two-stage process**.

The two stages are:

1. The light reaction
2. Carbon fixation

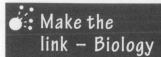

Make the link – Biology

Enzymes are biological catalysts (Chapter 5 page 60) They are required for photosynthesis. Photosynthesis is carried out by producers and provides the energy for food chains/food webs (Unit 3 pages 259–261).

Biology in context

Using immobilised algae is an interesting way to investigate photosynthesis. Algal balls can be made from a mixture of algae and sodium alginate (a gum extracted from brown algae). They will carry out photosynthesis under the right conditions.

The algal balls are placed into vials containing water and bicarbonate indicator. The vials are then placed under various light intensities ranging from darkness to bright light, along with a control where no indicator is used.

The bicarbonate indicator solution changes colour as the pH changes due to changes in carbon dioxide concentration during photosynthesis. If the concentration of carbon dioxide increases above 0.04%, the solution changes colour from red to yellow. If the carbon dioxide concentration decreases below 0.04%, it changes colour from red to purple. The colour changes allow deductions to be made about how light levels influence photosynthesis (fig 7.9).

Fig 7.9 *Experiment using immobilised algae*

Photosynthesis – Stage 1: The light reaction

The light reaction is the first stage of photosynthesis and takes place inside chloroplasts. Chloroplasts are organelles found inside plant cells and contain the pigment chlorophyll.

The number of chloroplasts found in a plant varies depending on the type of plant and the number of leaves.

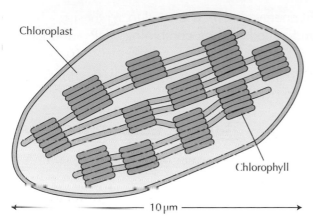

Fig 7.10 *Chloroplast structure*

The diagram of the chloroplast shows where the chlorophyll is found in the chloroplast (fig 7.10).

Light energy from the sun is trapped by **chlorophyll** in the **chloroplasts** and is converted to chemical energy.

Some of this chemical energy is used to split **water** into **hydrogen** and **oxygen**, a process also known as photolysis: photo meaning 'light' and lysis 'to split'. The splitting of water releases hydrogen, which attaches to hydrogen acceptor molecules and is transferred to the carbon fixation stage of photosynthesis. The excess oxygen diffuses out of the cell and is released into the atmosphere.

The light reaction is summarised in figure 7.11.

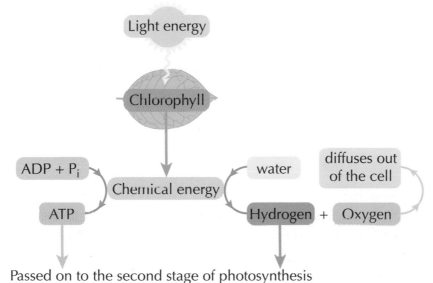

Passed on to the second stage of photosynthesis

Fig 7.11 *Stage 1: The light reaction*

Some of the light energy absorbed is used to synthesise a high-energy compound called ATP by combining ADP + inorganic phosphate (P_i).

ATP is required to provide energy in carbon fixation – the second stage of photosynthesis.

Photosynthesis – Stage 2: Carbon fixation

Carbon fixation is the second stage of photosynthesis, and it also takes place inside the chloroplast.

Energy from **ATP** is used to join **hydrogen** with **carbon dioxide**. Following a series of enzyme-controlled reactions, this produces the sugar glucose.

Pores called stomata (fig 7.12) are found on the lower leaf surface. These open during the day to allow carbon dioxide to diffuse into the leaves of a plant. This allows the plant to photosynthesise efficiently, as the products of the light reaction are required for carbon fixation to take place.

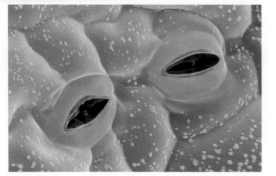

Fig 7.12 *The stomata on the underside of a leaf are visible when viewed under a high-powered microscope*

The carbon dioxide entering the leaf is combined with hydrogen from the light reaction to form **sugar** using energy in the form of ATP (also from the light reaction). As energy is released from the ATP molecule it is converted back into ADP + P$_i$ (the low-energy form of the molecule) (fig 7.13).

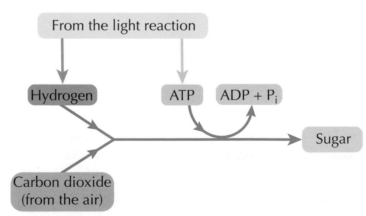

Fig 7.13 *Carbon fixation*

The summary diagram in figure 7.14 shows how the products of the light reaction are used in carbon fixation to produce sugar.

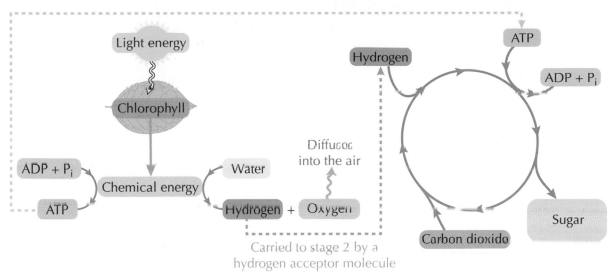

Fig 7.14 *Summary of stage 1 and stage 2*

GO! Activities

Activity 7.1 Working individually

(a) Write down full answers to the following questions:

1. Name the raw materials required for photosynthesis to take place.

2. Name the pigment that traps light energy from the sun.

3. Give the summary word equation for photosynthesis.

4. Name the two stages that make up the process of photosynthesis.

(b) Produce a leaflet giving detailed instructions of the procedure used to test leaves for starch. Include pictures.

Activity 7.2 Working individually

Write down full answers to the following questions:

1. Name the form of energy that light energy is converted to. Describe what it is used to do.

2. State what happens to excess oxygen produced during photosynthesis.

3. Name the products of the light reaction that are used in carbon fixation.

(Continued)

Make the link – Biology

Water is also lost through evaporation when the stomatal pores are open (Unit 2 page 179).

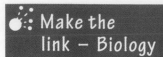

4. Describe what happens in the process of carbon fixation.

Activity 7.3 Working in pairs

Produce a poster to summarise the main two stages in photosynthesis.

The uses of sugar by plants

The sugar made in the process of photosynthesis is a source of chemical energy. This energy is available for **respiration** or can be converted into plant products such as **starch** and **cellulose**.

The uses of sugar by plants is summarised in the diagram below (fig 7.15).

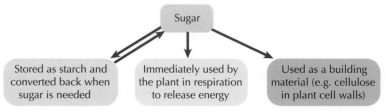

Fig 7.15 *The uses of sugar by plants*

Activity

Activity 7.4 Working individually

Produce a poster or leaflet to show the fates of the sugar produced during photosynthesis.

Unlike humans, plants cannot source different food types. The only food type available to plants is the carbohydrate (sugar) made during photosynthesis. The **carbohydrate** made by plants is then used to produce **fats** and **proteins**.

Limiting factors

When the sugar made during photosynthesis is used by the plant for respiration, this releases energy for growth. The use of sugar for growth allows the plant to grow new parts. The growth of new leaves allows the plant to carry out more photosynthesis and, in turn, make more food.

Factors such as **carbon dioxide concentration** and **light intensity** are essential for photosynthesis to take place. **Temperature** is also important for photosynthesis, as the process is controlled by enzymes, and enzymes are dependent upon temperature.

If any of these factors are in short supply, this will limit the rate of photosynthesis. For this reason, carbon dioxide concentration, light intensity and temperature are called **limiting factors**.

To find out how limiting factors affect photosynthesis, the rate of photosynthesis needs to be measured. This can be done by measuring one of the following over a set period of time: the volume of oxygen released by a plant; the carbon dioxide used by a plant; or the gain in dry mass of a plant.

Aquatic plants like *Elodea canadensis* (Canadian pondweed) and *Cabomba* are commonly used for measuring the rate of photosynthesis. If the plant stem is cut under water, the number of bubbles of oxygen released by the plant can be counted (fig 7.16).

The experiment shown in figure 7.17 is set up to investigate the effect of light intensity on the rate of photosynthesis.

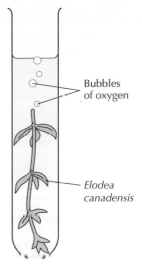

Bubbles of oxygen

Elodea canadensis

Fig 7.16 *Oxygen bubbles released by Elodea canadensis*

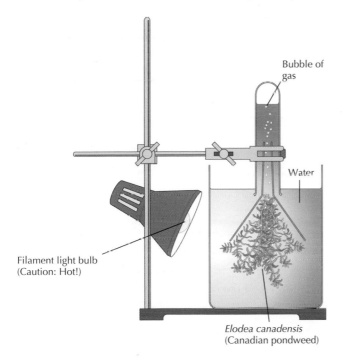

Bubble of gas

Water

Filament light bulb (Caution: Hot!)

Elodea canadensis (Canadian pondweed)

Fig 7.17 *Investigating light intensity*

The distance of the plant to the lamp is altered to vary the light intensity available to the plant for photosynthesis.

The graph in figure 7.18 shows the effect of increased light intensity on the rate of photosynthesis.

At point 1 on the graph, light intensity is a limiting factor. As the light intensity increases, so too does the rate of photosynthesis. At point 2, the light intensity has continued to be increased, but the rate of photosynthesis has not increased. Therefore, one of the other factors must be in short supply and is limiting the rate of photosynthesis.

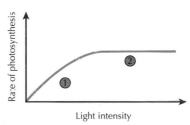

Rate of photosynthesis

Light intensity

Fig 7.18 *Effect of increased light intensity on the rate of photosynthesis*

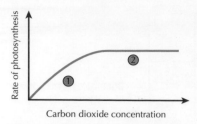

Fig 7.19 *Effect of carbon dioxide concentration on photosynthesis*

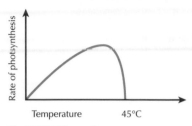

Fig 7.20 *Effect of temperature on photosynthesis*

The graph in figure 7.19 shows the effect of increased carbon dioxide concentration on the rate of photosynthesis.

At point 1 on the graph, carbon dioxide concentration is increasing, as is the rate of photosynthesis. This tells us that carbon dioxide concentration is the limiting factor. At point 2, the rate of photosynthesis stays the same despite the carbon dioxide concentration being increased. As the plant is being given more carbon dioxide, but the rate of photosynthesis is not increasing, the carbon dioxide concentration cannot be the limiting factor. One of the other factors must be in short supply.

The graph in figure 7.20 shows the effect of increased temperature on the rate of photosynthesis.

At low temperatures the rate of photosynthesis is slow. As temperature is increased, the rate of photosynthesis increases. The plant enzymes involved in photosynthesis work best between 20 and 25°C. As the temperature increases, the rate of photosynthesis begins to slow again, until 45°C where enzymes are denatured, and the rate of photosynthesis drops to zero.

Different factors can affect the rate of photosynthesis at different times. This is shown on the graph in figure 7.21.

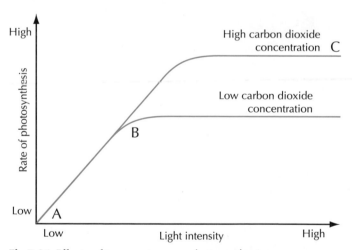

Fig 7.21 *Effects of temperature on photosynthesis*

Between points A and B on the graph (fig 7.21), light intensity is the limiting factor. Increasing the light intensity increases the rate of photosynthesis. After point B the rate of photosynthesis levels off, even though light intensity continues to increase. The limiting factor is no longer light intensity. Increasing the carbon dioxide concentration at point B causes an increase in the rate of photosynthesis, so it is concluded that carbon dioxide concentration is the limiting factor at point B.

At point C the rate of photosynthesis has again levelled off in spite of an increase in both light intensity and carbon dioxide concentration. The limiting factor must be temperature.

Only by eliminating the limiting factor can the rate of photosynthesis be increased. The more photosynthesis that takes place, the more sugar is produced. The more sugar there is, the more energy available to the plant for growth.

GO! Activity

Activity 7.5 Working individually

Read the information carefully, and then complete the tasks below.

In an experiment carried out to investigate the rate of oxygen production as light intensity is increased, the following results were obtained:

Light intensity (units)	Number of bubbles of oxygen produced in 2 minutes
0	0
5	12
10	21
15	34
20	45
25	45

(a) Draw a line graph of the results.

(b) Answer the following questions:

1. When light intensity is zero, why are no bubbles of oxygen produced?

2. Explain why the number of bubbles of oxygen produced can be used as a measure of the rate of photosynthesis.

3. When light intensity is increased from 20 to 25 units there is no increase in the number of bubbles of oxygen produced. Why is this?

4. What other factors could have had an effect on the number of bubbles of oxygen produced?

5. What could be done to improve the reliability of the results?

I can:

- State that photosynthesis is a series of enzyme-controlled reactions that occur in a two-stage process.

- State that the two stages in photosynthesis are: the light reaction (photolysis) and carbon fixation.

- Give the summary word equation for photosynthesis:

  carbon dioxide + water $\xrightarrow[\text{Chlorophyll}]{\text{Light energy}}$ sugar + oxygen

- State that light energy from the Sun is trapped by chlorophyll in the chloroplasts and converted into chemical energy in the form of ATP.

- State that ATP is used to split water (photolysis), producing hydrogen and oxygen.

- State that hydrogen attaches to hydrogen acceptor molecules.

- State that excess oxygen diffuses out of the cell.

- Describe how hydrogen and ATP produced by the light reaction are used in carbon fixation.

- Describe the process of carbon fixation as the joining of hydrogen and carbon dioxide to form sugar using ATP.

- State that the chemical energy in sugar can be used by the plant for respiration or converted to starch or cellulose.

- State that carbohydrate can be used to produce fats and proteins.

- State that carbon dioxide concentration, light intensity and temperature are limiting factors of photosynthesis.

- Explain that the absence or reduction of any one of the limiting factors reduces the rate of photosynthesis, slowing plant growth.

8 Respiration

Cellular respiration

Energy is required by all cells in the body for even the simplest of things. We need energy for movement, for keeping warm and even to keep our bodies working while we sleep.

Respiration is the chemical release of energy from the food that we eat. To ensure survival, all cells need to be able to carry out respiration.

To find out how much energy is present in different types of food, a simple experiment can be carried out. A food sample is burned in a set volume of water and the temperature increase measured. The energy contained in different food types can then be compared to determine which contains the most energy (fig 8.1).

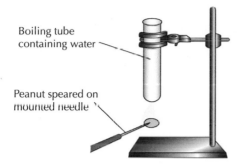

Boiling tube containing water

Peanut speared on mounted needle

Fig 8.1 *Measuring the energy content of food*

The chemical energy stored in **glucose** is released from cells through a series of **enzyme-controlled** reactions.

Respiration carried out in the presence of oxygen is called aerobic respiration. In the absence of oxygen it is called fermentation.

Make the link – Biology

Respiration should not be confused with breathing. Breathing is a mechanism that allows the exchange of gases. See Unit 2 page 201 for details of gas exchange in the lungs.

GO! Activity

Activity 8.1 Working individually

Write down full answers to the following questions:

1. Define 'respiration'.
2. Name the source of stored energy in cells.
3. The process of respiration involves a series of reactions. Describe how these reactions are controlled.

Make the link – Biology

Respiration is controlled by enzymes. Enzymes are affected by temperature (Chapter 5 page 65–66).

The release and use of energy

The energy released from the breakdown of glucose is used to generate ATP, a chemical store of energy. ATP is generated from ADP and phosphate during respiration.

ATP is formed when the chemical energy released during respiration is used to join together a molecule of ADP (adenosine diphosphate) with a single (P_i) group (figs 8.2 and 8.3).

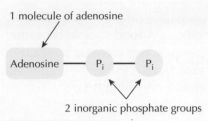

1 molecule of adenosine

2 inorganic phosphate groups

Fig 8.2 *ADP molecule*

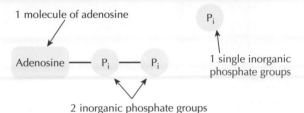

1 molecule of adenosine

1 single inorganic phosphate groups

2 inorganic phosphate groups

Fig 8.3 *The formation of ATP*

ATP is a high-energy molecule that consists of a molecule of adenosine joined to three inorganic phosphate (P_i) groups (fig 8.4).

Figure 8.4 summarises the link between energy and ATP.

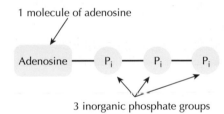

1 molecule of adenosine

3 inorganic phosphate groups

Fig 8.4 *ATP molecule*

The chemical energy stored in ATP can be released when required by cells by breaking it back down to ADP and inorganic phosphate. This releases the energy stored in the molecule, changing it to its low-energy state in the form of ADP + P_i.

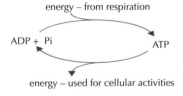

energy – from respiration

ADP + Pi   ATP

energy – used for cellular activities

The energy generated during respiration can be used for cellular activities including:

- muscle cell contraction
- cell division
- protein synthesis
- transmission of nerve impulses

Activity 8.2 Working individually

(a) Draw a simple diagram to show the link between ATP, the energy from respiration and the energy used for cellular activities.

(b) Draw a spider diagram to show the uses of energy.

Aerobic respiration

Respiration involves a series of enzyme-controlled reactions to completely break down glucose. This requires aerobic conditions: respiration in the presence of oxygen.

The basic word equation for aerobic respiration can be summarised as:

Glucose + oxygen → carbon dioxide + water + energy (38 ATP)

This takes places in a number of steps controlled by enzymes.

The first step in releasing the energy from glucose involves the breakdown of each glucose molecule to a substance called pyruvate. This first stage is also called glycolysis. It takes place in the **cytoplasm** of the cell and does not require oxygen. Two ATP molecules are formed as a result of breaking glucose down to pyruvate (fig 8.5).

The pyruvate must be further broken down to produce carbon dioxide and water. This stage requires oxygen and takes place in the **mitochondria**. During this stage, 36 ATP molecules are produced (fig 8.6).

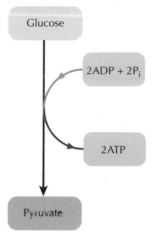

Fig 8.5 *Stage 1: The breakdown of glucose*

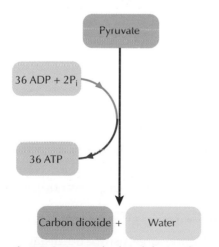

Fig 8.6 *Stage 2: The breakdown of pyruvate*

The complete aerobic breakdown of one glucose molecule in the presence of oxygen results in a yield of **38** molecules of ATP.

A summary of the whole process is shown in figure 8.7.

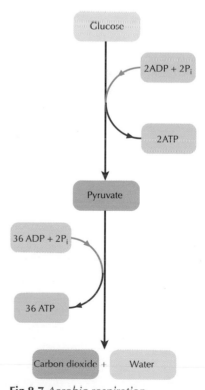

Fig 8.7 *Aerobic respiration*

GO! Activities

Activity 8.3 Working individually

Write down full answers to the following questions:

1. State the number of ATP molecules produced by the first stage of aerobic respiration.
2. State the number of ATP molecules produced by the second stage of aerobic respiration.
3. Name the final products that result from the complete breakdown of glucose.
4. Give the summary word equation for aerobic respiration.

Activity 8.4 Working in pairs

Using mini white boards or paper, take turns to practise drawing a diagram to show the complete breakdown of glucose in the presence of oxygen. Check each other's work.

Measuring the rate of respiration in animals

The rate of respiration can be measured using a respirometer. In the simple respirometer shown in figure 8.8, a woodlouse is placed inside a sealed tube.

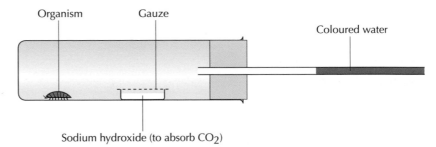

Organism   Gauze

Coloured water

Sodium hydroxide (to absorb CO_2)

Fig 8.8 *A simple respirometer*

The only air available to the organism is inside the tube. A solution of potassium hydroxide is used to absorb any carbon dioxide produced by the woodlouse. This way, only the oxygen used up by the organism is measured. As the organism uses up oxygen, the volume of gas inside the tube decreases. The coloured water moves down the tube to fill the space of the oxygen that has been used up by the organism. The distance that the liquid moves in a set time period is used to calculate the rate of respiration.

Respiration in germinating seeds

Germination is the growth of a seed. To enable seeds to grow they must carry out respiration. This allows the food stored within the seed to be broken down to release energy. This energy is then used for the growth of a seed into a plant.

To prove that seeds carry out respiration the following experiment is set up. The chemical limewater is included in the experiment, as it changes from clear to cloudy in the presence of carbon dioxide (fig 8.9).

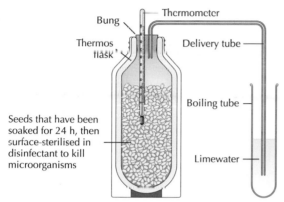

Fig 8.9 *Germinating seeds*

At the start of the experiment the temperature inside the flask is 20°C, and the limewater is clear. After 3 days the temperature has risen by 3 °C, and the limewater is cloudy.

GO! Activity

Activity 8.5 Working individually

Using the information given about the experiment described above complete the following tasks.

(a)

1. Construct a table to present the results of the experiment.
2. Explain how the results support the theory that the seeds are carrying out respiration.
3. Design a control experiment to show that it is the seeds that have caused the described changes to the limewater and temperature.

(b) Investigate your own rate of respiration by measuring your breathing rate before and after exercise.

*   While at rest, count the number of breaths taken in 1 minute.

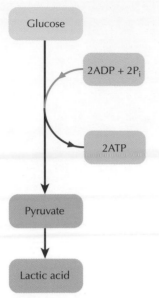

Fig 8.10 *Fermentation in animal cells*

- Do 3 minutes of exercise, e.g. jogging on the spot, star jumps or stepping up and down.
- Immediately count the number of breaths taken in 1 minute.
- Draw a bar chart of your results.

Answer the following questions:

1. What could you do to make the results more reliable?
2. Would another person get exactly the same results?

Fermentation in animal cells

In the absence of oxygen, respiration cannot take place as it normally would. Therefore, glucose cannot be completely broken down to carbon dioxide and water as it would when oxygen is present.

The **fermentation pathway** is followed and takes place without oxygen.

The first stage of glycolysis takes place as normal, as it does not require oxygen. Glucose is broken down to pyruvate, and two molecules of ATP are produced. In the absence of oxygen, pyruvate is then converted to **lactic acid**, as shown in figure 8.10.

This pathway is followed by animal cells when their demand for oxygen cannot be met by the body. This happens frequently during strenuous exercise.

When a person first starts to exercise the body will respire aerobically. As the intensity of the exercise increases, the breathing rate increases to a point where gas exchange cannot take place any faster. Energy is still required by the body cells, so respiration proceeds without oxygen. This allows energy to be released, but as a result lactic acid builds up in the muscles.

Eventually, the muscles tire, and exercise has to stop. Following a period of recovery where oxygen is available, the lactic acid is converted back to pyruvate.

Fermentation in animal cells can be summarised in the following word equation:

Glucose → little energy (2 ATP) + lactic acid

🔍 Hint

Think of lactic acid as an oxygen 'debt'. Lactic acid is allowed to build up in the muscles when there is a lack of oxygen. As soon as oxygen is available it is 'repaid', and the lactic acid is removed.

Make the link – PE, Health and Wellbeing

The demand for energy increases during exercise. This increases the demand for oxygen. This is why breathing rate and depth increase during exercise.

Activity 8.6 Working individually

Write down full answers to following questions:

1. Describe the process of fermentation in animal cells.
2. State the number of ATP molecules produced during fermentation in animal cells.
3. Name the final product that results from fermentation in animal cells.
4. Give the summary word equation for fermentation in animal cells.

Fermentation in plant and yeast cells

In plant and yeast cells a similar situation arises. The fermentation pathway is followed. Glycolysis takes place without oxygen, and glucose is broken down to pyruvate. Two molecules of ATP are produced as a result of this process.

The difference between animal cells and plant and yeast cells occurs in the next stage. In plant and yeast cells, pyruvate is broken down to **ethanol** (alcohol) and **carbon dioxide** as shown in figure 8.11.

As the carbon dioxide produced is released the reaction cannot be reversed.

Fermentation in plant and yeast cells can be summarised by the following word equation:

Glucose → little energy (2 ATP) + ethanol + carbon dioxide

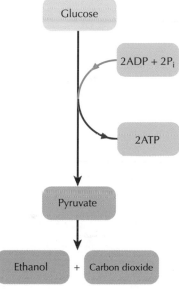

Fig 8.11 *Fermentation in plant and yeast cells*

Activity 8.7 Working individually

Write down full answers to the following questions:

1. Describe the process of fermentation in plant and yeast cells.
2. State the number of ATP molecules produced during fermentation in plant and yeast cells.
3. Name the final products that result from fermentation in plant and yeast cells.
4. Give the summary word equation for fermentation in plant and yeast cells.

🌳 Biology in context

The process of fermentation is used commercially in industry.

Yeast cells are used in the brewing industry to produce alcohol in beer and wine.

Yeast is also used in bread making, as the production of carbon dioxide makes the dough rise (fig 8.12).

Fig 8.12 *Products of fermentation*

Location of aerobic respiration and fermentation

The process of aerobic respiration starts in the **cytoplasm** and is completed in the **mitochondrion**. The mitochondrion is a sausage-shaped organelle found in the cytoplasm of a cell.

The detail of a mitochondrion as revealed using a high-powered microscope is shown below (fig 8.13).

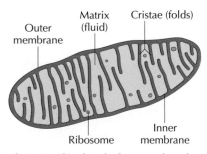

Fig 8.13 *The detail of a mitochondrion*

Muscle, companion, sperm, and nerve cells all have many mitochondria present in the cytoplasm of their cells as they require a lot of energy.

The process of **fermentation** is completed in the cytoplasm. The breakdown of each glucose molecule via the fermentation pathway yields 2 molecules of ATP when oxygen is not present.

☉☉ Activities

Activity 8.8 Working individually

(a) Copy and complete the table below to compare aerobic respiration and fermentation.

	Aerobic respiration in all cell types	Fermentation in	
		Animals	Plants and yeast
Products (s)			
Number of ATP molecules produced (per molecule of glucose)			

(b) Choose a cell type that contains many mitochondria. Produce an information leaflet on the uses of energy in your chosen cell.

You should include:

- a drawing of the cell showing many mitochondria present
- an explanation of how energy released from respiration is used to build up and break down ATP.

Activity 8.9 Working with a partner

Produce a poster to compare aerobic respiration and fermentation. Divide the poster into three sections: one for aerobic respiration, one for fermentation in animal cells, and the final section for fermentation in plant and yeast cells.

You should include:

- as many differences between the processes as possible
- the names of the parts of the cell where each process takes place
- diagrams where possible
- word equations for all processes.

I can:

- Explain that respiration is a series of enzyme-controlled reactions carried out by all cells to release the chemical energy stored in glucose.

- Describe how the energy released from the breakdown of glucose is used to generate ATP from ADP and inorganic phosphate.

- Describe how the chemical energy stored in ATP can be released by breaking it down to ADP and inorganic phosphate.

- State that ATP can be regenerated during respiration.

- Name the cellular uses of ATP (to include muscle-cell contraction, cell division, protein synthesis and transmission of nerve impulses).

- Describe the breakdown of glucose via pyruvate to carbon dioxide and water.

- State that aerobic respiration yields 38 molecules of ATP.

- Give the summary word equation for respiration:

     Glucose + oxygen → carbon dioxide + water

- Describe the breakdown of glucose via pyruvate to lactic acid in animal cells.

- Give the summary word equation for fermentation in animal cells:

     Glucose → lactic acid

- Describe the breakdown of glucose via pyruvate to alcohol/ethanol and carbon dioxide in plant and yeast cells.

- Give the summary word equation for fermentation in plant and yeast cells:

     Glucose → ethanol/alcohol + carbon dioxide

- State that aerobic respiration starts in the cytoplasm and is completed in the mitochondria.

- State that cells such as muscle, companion, sperm and nerve cells have a high number of mitochondria, as they require a lot of energy.

- State that fermentation occurs in the cytoplasm.

- State that the breakdown of each glucose molecule via the fermentation pathway yields two molecules of ATP when oxygen is not present.

Unit 1– Assessment

Section A

1. Which two structural features are common to both plant and bacterial cells?

   A Cell wall and ribosomes

   B Cell wall and nucleus

   C Plasmid and ribosomes

   D Sap vacuole and nucleus

2. Which of the following are two components of the cell membrane?

   A lipids and carbohydrates

   B Proteins and fats

   C Carbohydrates and proteins

   D Proteins and lipids

3. Typical timings of the stages of mitosis are shown in the table below.

Stage	1	2	3	4
Time (minutes)	87	34	26	53

   What percentage of the total time for mitosis is taken by stage 3?

   A 13

   B 26

   C 37

   D 74

4. The diagram below shows a section of unwound DNA.

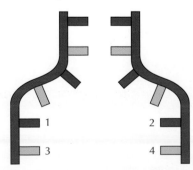

If Number 1 represents the base G, then number 2 must represent the base

A   G

B   A

C   T

D   C

5. The DNA of a chromosome carries information that determines the structure and function of

A   base

B   fats

C   proteins

D   carbohydrates

6. The active site of an enzyme is complementary to

A   one type of product molecule

B   all types of product molecules

C   one type of substrate molecule

D   all types of substrate molecules

7. The steps involved in the process of genetic engineering are shown below:

1   insert required gene into vector/bacterial plasmid

2   extract required gene from source chromosome

3   grow transformed cells to produce a genetically modified organism

4   insert plasmid into host cell

5   identify section of DNA that contains required gene from source chromosome

The correct order for these steps is

A   3, 5, 2, 4, 1

B   2, 5, 1, 4, 3

C   5, 2, 1, 4, 3

D   1, 4, 3, 2, 5

8. The raw materials for photosynthesis are

A   carbon dioxide and water

B   oxygen and water

C   carbon dioxide and sugar

D   oxygen and sugar

9. The graph below shows the effect of light intensity, concentration of carbon dioxide and temperature on the rate of photosynthesis.

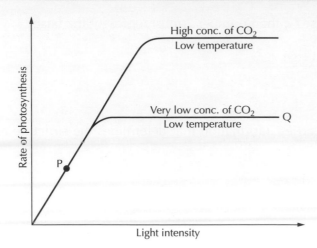

Which factors are limiting the rate of photosynthesis at points P and Q on the graph?

	P	Q
A	Light	Carbon dioxide
B	Light	Temperature
C	Carbon dioxide	Temperature
D	Temperature	light

10. Which line in the table identifies correctly the location and products of fermentation in plant cells?

	Location	Products
A	Mitochondria	Water + carbon dioxide + 38 ATP
B	Cytoplasm	Ethanol + carbon dioxide + 2 ATP
C	Mitochondria	Ethanol + carbon dioxide + 2 ATP
D	Cytoplasm	Water + carbon dioxide + 38 ATP

Section B

1. The diagram below shows a yeast cell.

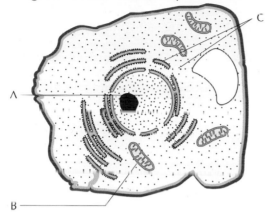

(a) Complete the table below to give the name and function of the parts labelled A, B and C.

Letter	Structure	Function
A	Nucleus	
B	Mitochondrion	
C		Site of protein synthesis

2

(b) Name the structure that is present in plant cells but is absent from yeast cells.

_____ 1

2. The diagram below represents the transfer of oxygen from the atmosphere to muscle cells.

Atmosphere ⟶ Red blood cells Muscle cells

(a) From the diagram, state where the highest and the lowest concentrations of oxygen are found.

Highest concentration _____

Lowest concentration _____

1

(b) Name the process by which oxygen enters cells.

_____ 1

3. Describe the osmotic effect of transferring:

Animal cells from a weak salt solution to a solution of pure water.

or

Plant cells from a weak sugar solution to a strong salt solution.

_____

_____

_____

_____

_____

3

4. (a) The diagram below contains some of the stages of mitosis.
Describe **Stages 2** and **5** in the spaces provided.

Stage 1
Chromosomes become visible as pairs of identical chromatids

↓

Stage 2

1

↓

Stage 3
The spindle fibres shorten, pulling the chromatids to the opposite poles of the cell

↓

Stage 4
The nuclear membrane re-forms around each group of chromatids

↓

Stage 5

1

(b) Cells can be produced in the lab by cell culture. This requires the use of aseptic techniques.

Describe two precautions that should be taken to ensure aseptic conditions.

Precaution 1

_____

_____

1

Precaution 2

_____

_____

1

5. The diagram below shows how a protein chain is assembled from the DNA code.

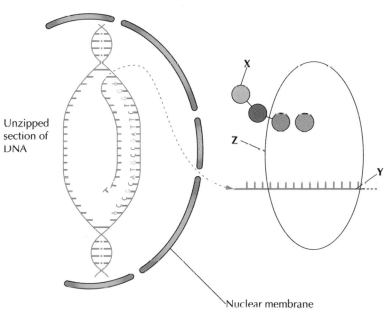

(a) Name molecules X and Y.

X _____

Y _____

1

(b) Name organelle Z where the protein chain is being assembled.

_____

1

6. The diagram below represents the molecular structure of the enzyme catalase.

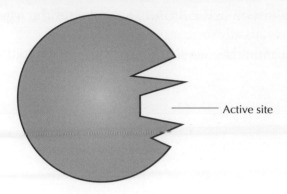

Active site

What happens to the active site when an enzyme is denatured?

_____

1

7. Three groups of students investigated the catalase concentration of different tissues.

Each group set up a test tube containing 5 cm³ of hydrogen peroxide and a cube of tissue. The oxygen was collected over a 5 minute period and the volume was measured as shown in the diagram below.

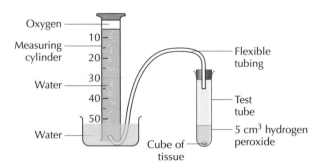

The procedure was repeated by each group using cubes of liver, apple, carrot and plasticine.

The results from the three groups are given in the table below.

Tissue	Volume of oxygen collected in 5 minutes (cm³)			
	Group 1	Group 2	Group 3	Average
Liver	44.5	43.5	44	44.0
Apple	1.0	1.5	0.5	1.0
Carrot	4.5	3.0	3.0	3.5
Plasticine	0	0	0	0

(a) The volume of hydrogen peroxide and time taken to collect the oxygen were kept constant in this investigation.

State two other variables that must be kept constant.

1 _____ 1

2 _____ 1

(b) Why was plasticine included as a control in this experiment?

_____

_____ 1

(c) What was done in this investigation to make the results reliable?

_____

_____ 1

(d) What conclusion can be drawn from these results?

_____

_____ 1

8. The graph shows the effect of temperature on the enzyme amylase.

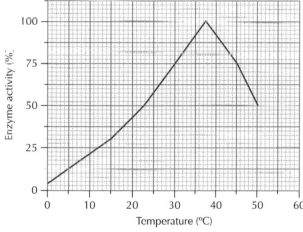

(a) Between which two temperatures was there the greatest overall increase in enzyme activity?

Tick (✓) the correct box

0°C to 10°C ☐
10°C to 20°C ☐
20°C to 30°C ☐
30°C to 40°C ☐   1

(b) From the graph, predict the temperature at which the enzyme activity will reach zero

_____°C

1

9. An experiment was set up to measure the effect of light intensity on the rate of photosynthesis in the aquatic plant, *Elodea*.

The light intensity was varied using a dimmer switch on the bulb.

The rate of photosynthesis was measured by counting the number of bubbles released per minute.

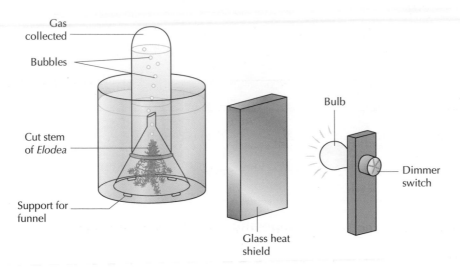

The results of the experiment are shown in the table below.

Light intensity (units)	Rate of photosynthesis (number of bubbles per minute)
2	4
4	12
6	28
8	48
10	48
12	48

(a) On the grid below, plot a line graph to show the rate of photosynthesis against light intensity.

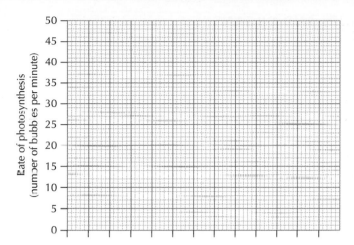

2

(b) There are two stages in photosynthesis.

Hydrogen and a high-energy molecule are produced during the light reaction.

(i) Name the high-energy molecule.

_____ **1**

(ii) Describe the use of hydrogen in carbon fixation.

_____ **1**

10. The diagram below shows aerobic respiration in an animal cell.

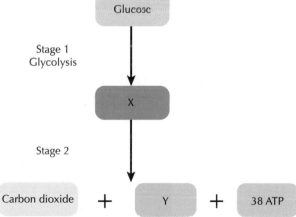

(a) Name substances X and Y.

X _____

Y _____ **2**

(b) Name the part of the cell where stage 1 takes place.

_____ **1**

(c) Describe how the reaction would differ in the absence of oxygen.

_____ **1**

UNIT 2
Multicellular Organisms

9 Cells, tissues and organs

You should already know:

- Cells can have different structures that allow them to carry out different functions.
- The structure and function of key organs and organ systems. For example, the circulatory system allows blood to be pumped around the body and involves organs such as the heart and blood vessels.

Learning intentions

- Explain what is meant by the term 'specialisation of cells in animals and plants'.
- Give details of how the structure of an animal cell or plant cell can relate to its function.
- Describe the levels of organisation found in animals and plants.
- State that the cells in organs are specialised for a specific function.

Specialisation in animal cells

Animal cells show **specialisation**. This means they have a **special shape or structure** that allows them to carry out a **specific function**. The table below shows some animal cells and explains how their structure is related to their function.

Cell	How cell structure relates to function
	Red blood cells carry oxygen around the body. They contain haemoglobin to do this. Red blood cells do not contain a nucleus or ribosomes, so they can contain as much haemoglobin as possible. Red blood cells have a special biconcave shape, which increases their surface area so they can absorb oxygen quickly and efficiently.
	Sperm cells swim towards the egg cell to carry out fertilisation. They have a tail to enable them to swim and many mitochondria for energy.

Cell	How cell structure relates to function
	Muscle cells contract and relax to allow movement. They are long and thin, and contain special contractile proteins called myofilaments, which allow them to become shorter, creating muscle contraction.
	Neurons send messages around the body as electrical impulses. They are long and thin, and have many projections that allow them to communicate with many other neurons.
	An **egg cell** fuses with a sperm cell at fertilisation and divides many times. They have cytoplasm containing large amounts of food to allow them to do this.

Levels of organisation

Animals have levels of organisation in their bodies that allow them to function efficiently. Specialised cells that perform specific functions group together to form **tissues**. For example, muscle cells group together to form muscle tissue. Tissues work together to form **organs** with specific functions. An example is the heart which contains muscle and nervous tissue. The cells in organs are specialised for their function. For example, the muscle cells in the heart are able to contract in a regular pattern. This allows the heart to beat and pump blood around the body. Organs work together forming a **system**. The heart is part of the circulatory system, which also includes the blood vessels (fig 2.1).

Make the link – Biology

Body cells have many different functions (see pages **114, 127, 186**)

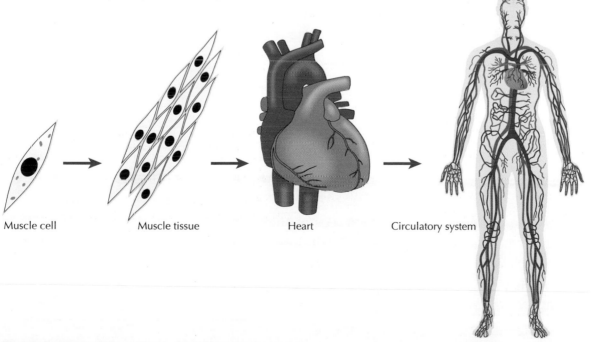

Muscle cell   Muscle tissue   Heart   Circulatory system

Fig 2.1 *The levels of organisation found in animals*

GO! Activities

Activity 2.1.1 Working individually

1. Explain the term 'specialisation'.
2. Describe how a red blood cell is specialised for its function.
3. Put the following structures in order from smallest to largest:

    cell   system   organ   tissue

4. Complete the table below with three examples of systems in animals.

System	Function	Organs in this system	Cells in this system
Cardiovascular system			
Nervous system			
Female reproductive system			

Activity 2.1.2 Working individually

Make an 'Animal cell owner's manual'. Cut two sheets of A4 paper in half and arrange them into a booklet, stapling the centre together. On each page of the booklet draw a diagram of a different type of animal cell and explain how its structure is related to its function.

I can:

- Describe specialisation of animal cells as a process whereby cells develop a specific function to which they are suited.

- State that animal cells have a structure that allows them to carry out their function and give an example.

- Describe the levels of organisation seen in animals from groups of cells forming a tissue, different tissues forming organs and organs working together to form a system.

- State that the cells in organs are specialised for their function.

Specialisation in plant cells

Like animal cells, plant cells show **specialisation**. Plant cells are adapted to carry out many different functions. The table below shows some plant cells and explains how their structure is related to their function.

Cell	How cell structure relates to function
	Root hair cells absorb water. They have a large projection that increases their surface area for absorbing water.
	Palisade mesophyll cells are found in leaves. They contain a large number of chloroplasts to allow them to carry out photosynthesis efficiently.
	Xylem cells transport water in plants. Xylem cells take the form of hollow tubes that allow water to travel from the roots to the leaves in a continuous stream.
	Phloem cells transport food in plants. The end walls of phloem cells are like sieves, which allow glucose to pass from one cell to the next.
	Plant leaves have pores called stomata, which allow gases to enter and leave the leaf. **Guard cells** change shape to open or close the pore.

Make the link – Biology

Plant cells have many different functions (see pages **117, 150, 176** and **183**)

Levels of organisation in plants

Like animals, plants have different levels of **organisation**. For example xylem **cells**, which transport water, join together to form xylem **tissue**. Xylem tissue works alongside phloem tissue in an **organ** called the vascular bundle. The cells in plant organs are specialised to carry out a specific function. For example the phloem cells in the vascular bundle have end walls like sieves that allow sugar to pass easily from one cell to the next. The vascular bundle is part of the vascular **system** found in plants.

GO! Activities

Activity 2.1.3 Working individually

1. Copy and complete the diagram below to show the levels of organisation found in plants.

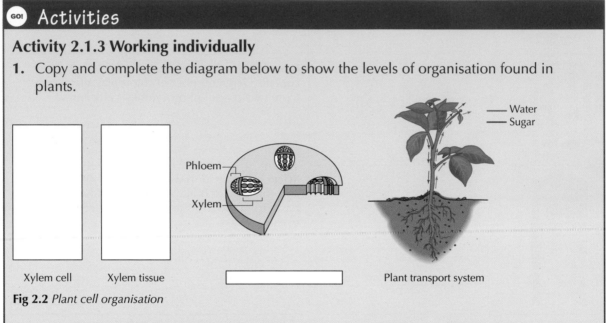

Fig 2.2 *Plant cell organisation*

2. Describe how a palisade mesophyll cell is specialised for its function.

Activity 2.1.4 Working individually

Make a 'Plant cell owner's manual'. Cut two sheets of A4 paper in half and arrange them into a booklet, stapling the centre. On each page of the booklet draw a diagram of a different type of plant cell and explain how its structure is related to its function.

I can:

- Describe specialisation of plant cells as a process whereby cells develop a specific function to which they are suited.

- State that plant cells have a structure that allows them to carry out their function and give an example.

- Describe the levels of organisation seen in plants from groups of cells forming a tissue, different tissues forming organs and organs working together to form a system.

- State that the cells in organs are specialised for their function.

10 Stem cells and meristems

Make the link – Biology

Specialisation involves a cell becoming suited to a particular function or role (see page **114**).

Stem cells

Stem cells are found in animals. They can divide to produce more stem cells or can develop into specialised cells. Stem cells have the ability to turn into any type of body cell and are involved in **growth** and **repair** of body tissues. Scientists are currently conducting medical research and developing treatments for certain illnesses that involve the use of stem cells. There are two main types of stem cells: embryonic stem cells and adult stem cells.

Embryonic stem cells are found in embryos. Stem cells obtained from embryos have the ability to develop into any type of body cell. They allow an organism to grow from a tiny embryo to a fully formed individual. Research on the possible use of embryonic stem cells in medicine is underway but there are ethical issues involved.

Adult stem cells are found throughout the body. Despite being called adult stem cells, they can be found in children as well as adults. Adult stem cells do not have the same properties as embryonic stem cells. They are slightly more specialised than embryonic stem cells. For example certain stem cells found in the bone marrow are only capable of becoming a variety of different blood cell types.

Stem cells have important **medical applications**. The table on page 121 shows the current uses of stem cells in medicine.

Source of stem cell	Medical use
Marrow in the centre of bones	Treating leukaemia, a type of cancer caused by abnormal blood cells.
Skin	Growing new layers of skin that can be used to treat burn victims.
Heart muscle	Repairing damaged heart muscle after a heart attack (this technique is still being tested).
Bladder	Building a new bladder in a laboratory for a patient whose bladder has been damaged by injury or disease.

Fig 2.3 *Cross section through a bladder. The bladder is a hollow organ and is therefore easier to build than an organ such as the heart*

In the future, scientists hope that stem cells will be used to cure conditions such as Alzheimer's disease, diabetes, spinal-cord injuries and stroke. Stem cells may also provide a useful alternative to animals for testing experimental drugs.

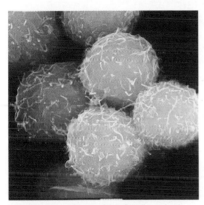

Fig 2.4 *A coloured high-powered microscope image of stem cells*

Fig 2.5 *Skin grown in a lab from stem cells is sliced and stretched before being used to treat burn victims*

🌳 Biology in context

There are many issues surrounding the use of stem cells, in particular embryonic stem cells.

• Research using embryonic stem cells involves the destruction of embryos. Many people believe that it is unethical to destroy embryos because they have the potential to grow into a baby.

• Some people believe we should use only adult stem cells because these do not require the destruction of embryos.

• Some of the techniques used in stem-cell research are closely related to cloning, which many people oppose.

• Like any new medical advance, there are also safety concerns surrounding the use of stem cells. For example, some research has suggested that transplanted stem cells could become cancerous.

GO! Activities

Activity 2.2.1 Working individually

Type the following address into your browser. Watch the clip and answer the questions below.

www.bbc.co.uk/learningzone/clips/stem-cell-research/6013.html

1. What do the scientists hope stem cells will enable them to do?
2. What is the easiest way to obtain stem cells at the moment?
3. What are stem cells?
4. Why are stem cells special?
5. How could stem cells be used to help the woman shown in the clip?
6. What are the two main ethical issues surrounding the use of stem cells?

Activity 2.2.2 Working individually

Carry out some research into stem cells obtained from embryos. Write an essay that discusses the use of embryonic stem cells. Your essay should be laid out in the following format:

Introduction – introduce the issue

Arguments for – give two or three arguments for using stem cells obtained from embryos

Arguments against – give two or three arguments against using stem cells obtained from embryos

Conclusion – summarise your essay and give your own opinion.

I can:

- Describe the role of stem cells in animals as the cells that give rise to specialised cells.

- State that stem cells have the potential to divide to become different types of cells.

- State that stem cells are involved in growth and repair.

Meristems

In animals, growth can occur all over their bodies. In plants, growth is restricted to special points called **meristems**. Meristems can be found at the tips of the shoots in plants. They can also be found at the tips of the roots, protected by a group of cells called the root cap. **Cell division** takes place in meristems and produces **non-specialised** cells. These cells have the potential to become any type of plant cell. Cell division at meristems allows plants to **grow**.

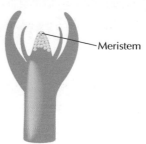

Fig 2.6 *Meristems can be found at the tip of the shoots in plants*

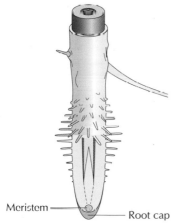

Fig 2.7 *Meristems can be found at the tip of the roots in plants*

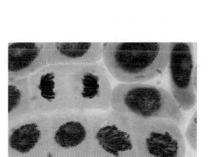

Fig 2.8 *The root tip of garlic can be stained to show mitosis taking place*

GO! Activities

Activity 2.2.3 Working individually

1. Draw a simple diagram to show the location of the root and shoot meristems in a plant. Label the meristems in your diagrams.

2. Name the process that takes place in meristems.

3. Describe how plant growth differs from animal growth.

4. Explain how the process of mericloning involves the use of meristems.

⚫: Make the link – Biology

Mitosis is the process of nuclear division that usually comes before cell division (see page **41**)

I can:

- State that meristems are the only place where cell division occurs in plants.

- State that cells produced in meristems are non-specialised and have the ability to become any type of plant cell.

- State that growth occurs due to the activity of meristems in plants.

11 Control and communication

You should already know:

- The structure and function of key organs and organ systems.
- For example the nervous system, which allows messages to be sent to and from the brain.

Learning intentions

- Describe the structure of the brain.
- Give details of the functions of the different structures of the brain.
- Describe the structure and function of the central nervous system (CNS).
- State that the CNS is a communication system.
- Describe the role of sensory and motor neurons.
- Describe the different responses brought about by the CNS.
- Explain what a reflex action is and give some examples.
- Explain the importance of reflex actions.
- Explain the role of sensory, relay and motor neurons in a reflex arc.
- State that electrical impulses move along neurons.
- Explain the importance of synapses in the nervous system.
- State the function of endocrine glands.
- Describe the role of hormones in signalling.
- Describe the role of the liver, pancreas, insulin and glucagon in controlling blood glucose levels.
- State that glucose can be stored as glycogen in the liver.
- Describe the causes and treatment of type 1 and type 2 diabetes.
- Explain the importance of controlling blood glucose levels.

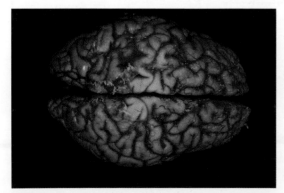

Fig 2.9 *The outer surface of the brain seen from above*

The brain

The brain is the most complex organ in your body. It controls vital processes that keep you alive, enables you to interact with the world around you and makes memories that you can keep for decades. Your brain is capable of doing several things at once. For example right now you are reading, sitting upright, breathing, and your heart is beating. All of these processes require input from your brain.

The brain is organised into distinct areas, each of which carries out a different function. The largest part of the brain is the **cerebrum**. The cerebrum is the area of the brain that enables conscious thought and memory. The cerebrum also receives information from the senses, processes the information and brings about a response. The **cerebellum** is found at the back of the brain, tucked underneath the cerebrum. The cerebellum is responsible for controlling balance and co-ordination. The **medulla** lies above the spinal cord. The medulla is the region of the brain that controls breathing and heart rates.

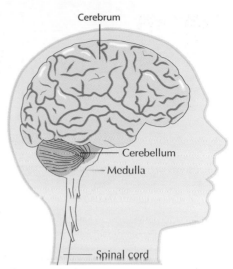

Fig 2.10 *The brain*

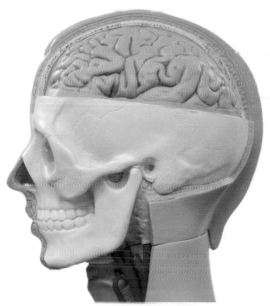

Fig 2.12 *The brain is protected by the skull*

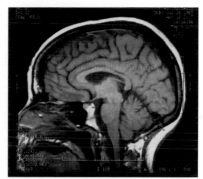

Fig 2.11 *A scan called an MRI can be used to produce an image of the brain*

🌳 Biology in context

Sleep is extremely important for the functioning of the brain. People who experience a period of time without sleep have difficulty concentrating and a reduced attention span. If a period of sleeplessness continues, memory, planning and language can also be affected. On average, adult humans require 8 hours of sleep per night. Other animals require varying amounts of sleep – from a giraffe, which needs just 2 hours, to a tiger, which needs around 16 hours.

Scientists are unsure exactly why we need to sleep but several theories have been put forward:

- Sleep allows the body to repair itself from damage caused through the day.
- Sleep allows the body to replenish its store of ATP.
- Sleep allows your brain to replay events of the day to cement them in your memories.

☄ Make the link – Expressive Arts

In art you practise still-life drawing. Early anatomists often used artists to record their work, before cameras were widely available.

☄ Make the link – Biology

The brain is part of the nervous system (see page **127**).

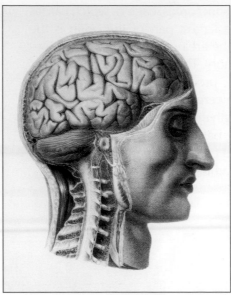

Fig 2.13 *This artwork was completed for an anatomy book published in the mid 1800s*

GO! Activities

Activity 2.3.1 Working individually

1. Copy and complete the table below into your notes.

Structure	Function
Cerebrum	
	Controls heart rate and breathing rate
Cerebellum	

Activity 2.3.2 Working in pairs

The 'Stroop effect' was named after a scientist called J. Ridley Stroop, who carried out this experiment in the 1930s. Look at the Stroop effect picture and try to name the colours of the different words. Do not read the word itself, try to name its colour. Try one set at a time.

Which set did you find easier?

Why might this be?

Would this test be easier for young children who know their colours but cannot read yet?

Red	Green
Blue	Red
Green	Blue
Black	Black
Green	Blue
Blue	Red

Fig 2.14 *The Stroop effect*

Activity 2.3.3 Working in pairs

There are two different types of sleep, non-REM and REM sleep. Research each type and write a 200-word report describing your findings.

I can:

- Describe and identify the structures of the brain as the cerebrum, cerebellum and medulla.
- State that the cerebrum enables conscious thought and memory.
- State that the cerebellum controls co-ordination and balance.
- State that the medulla controls heart and breathing rates.

The nervous system

The nervous system is made up of the brain, spinal cord and nerves. This is a communication system that uses electrical impulses to send messages from one part of the body to another. This allows communication between the cells of a multicellular organism. Neurons (nerve cells) send messages as electrical impulses from one part of the nervous system to another. Neurons are specially structured to allow them to send signals quickly and efficiently. The **CNS** is made up of the **brain** and the **spinal cord** only.

Sense organs contain **receptor cells** which detect stimuli (changes in conditions). This information is passed along **sensory neurons** to the CNS. The CNS processes the information and brings about responses. Responses are passed along **motor neurons** to **effectors**. An effector may be a muscle that brings about a rapid response or a gland that brings about a slower response.

One example of nervous-system communication is shown in figure 2.16. The nervous system brings about changes when there is a drop in the external temperature of the body. Shivering is a faster response because it is brought about by muscles. The reduction in sweating occurs more slowly because this is brought about by a gland.

✷: Make the link – Biology

A neuron is an example of a cell that is well suited to carrying out its function (see page **115**).

✷: Make the link – Sciences

In physics you study electrical charges. Electrical impulses allow messages to be sent around the body. Some organisms such as the electric eel use electrical charges for other reasons such as defence.

🔎 Hint

Remember the CNS is the brain and spinal cord only. Think about a diagram of the body facing you. The brain and spinal cord are in the centre.

Fig 2.15 *The nervous system*

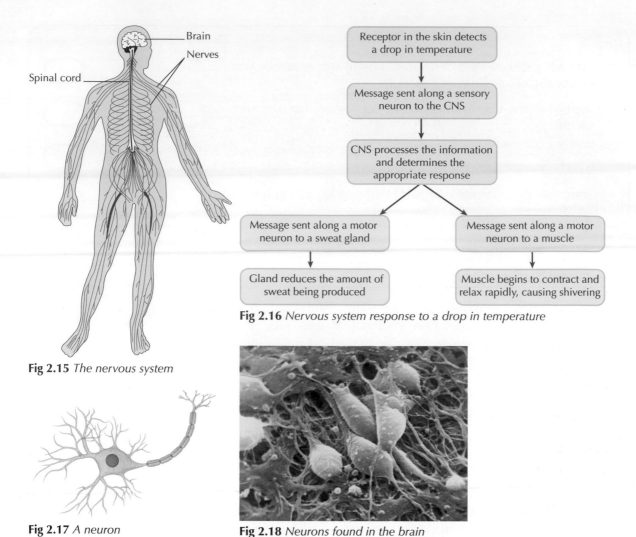

Fig 2.16 *Nervous system response to a drop in temperature*

Fig 2.17 *A neuron*

Fig 2.18 *Neurons found in the brain*

◉ Investigation

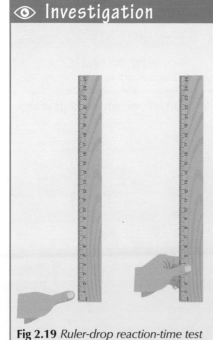

Reaction time measures the length of time it takes your nervous system to respond to a stimulus. This can be measured using a ruler drop test. The subject holds their thumb and fingers either side of a ruler. The person carrying out the test drops the ruler, and the subject tries to catch it as quickly as possible. Their reaction time is measured as the point on the ruler scale where their hand catches it. The shorter the distance, the faster the reaction time.

Planning and designing an investigation

Design an experiment that would allow you to investigate the effect of age on reaction time.

What factors would need to be kept the same in this investigation?

Making a prediction

How do you think age will affect reaction time? Write down your hypothesis.

Fig 2.19 *Ruler-drop reaction-time test*

Results

Here are some sample results from an experiment to investigate the effect of distraction on reaction rate.

Distraction level	Attempt 1 (distance on ruler scale, mm)	Attempt 2 (distance on ruler scale, mm)	Attempt 3 (distance on ruler scale, mm)	Average (distance on ruler scale, mm)
No distraction	195	175	155	
Quiet music	205	170	165	
Moderate music	200	180	190	
Loud music	230	215	215	

Calculate the average reaction time at each level of distraction.

Draw a bar graph to display the average results.

Conclusion

What do these results show?

What is the relationship between distraction and reaction rate?

Can you explain these results?

Evaluation

How could the reliability of this experiment be improved?

GO! Activities

Activity 2.3.4 Working individually

1. Draw a diagram of the nervous system and label the brain, spinal cord and nerves.
2. Name the two structures which make up the central nervous system.
3. Describe the role of sensory and motor neurons.
4. Copy and complete the sentence below, selecting the correct option from each bracket.

Effectors bring about responses. Muscles bring about (rapid/slow) responses whereas glands bring about (rapid/slow) responses.

Activity 2.3.5 Working in a group

Set up three bowls of water. The left bowl should contain cold water with ice cubes, the middle should be left at room temperature, and the right bowl should contain hot (not burning hot!) water. Place your left hand in the left bowl and right hand in the right bowl for a few minutes. Then move both hands into the middle bowl at the same time.

What do you feel?

Is there a difference between your left and right hands?

Why might this be?

(Continued)

Activity 2.3.6 Working in pairs

Carry out the ruler drop test as described above, using sight to test reaction time. Try it a second time with the subject blindfolded. The person carrying out the test should say 'Go' when they drop the ruler this time.

Was there a difference in reaction time when using sound rather than sight as a stimulus? Why might this be?

I can:

- State that the CNS is made up of the brain and the spinal cord.

- Describe the function of the CNS as processing information from the senses and bringing about appropriate responses.

- State that the CNS allows communication between the cells of a multicellular organism.

- State that a stimulus is a change in conditions and is detected by receptors.

- Describe the role of sensory neurons as passing information to the CNS.

- Describe the role of motor neurons as enabling a response to occur.

- State that responses brought about by the CNS may be fast, such as those from a muscle or slow, such as those from a gland.

Reflex action

Reflex actions are fast, automatic responses. They occur in response to a specific stimulus and usually have a **protective** role. Reflex actions happen without any input from the brain and therefore cannot be controlled.

For example, if your hand touches a very hot object, it will automatically pull away. This happens very **rapidly** to remove the hand from danger as quickly as possible. The table on the next page shows some examples of reflex actions. Reflex actions protect the body from harm. For example the gag reflex helps prevent choking. The pupillary light reflex protects the sensitive cells at the back of the eye from damage.

Name of reflex action	Stimulus	Effect	Advantage
Pupillary light reflex	Light intensity	Pupils become smaller in bright light	Cells at the back of the eye are protected from high light intensity
Blink reflex	Touching the eye	Eyelids blink	Protects the eye from potential damage
Gag reflex	Touching the back of the tongue or throat	Muscles in the back of the throat contract, causing gagging	Helps to prevent choking
Pharyngeal swallow reflex	Food touching the back of the throat	Causes swallowing	Helps to prevent choking
Mammalian diving reflex	Cold water touching the face	Heart rate slows	Allows survival underwater for longer periods of time

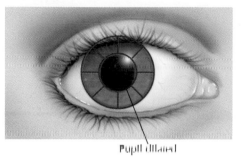
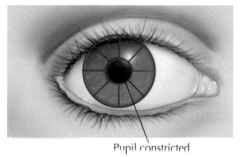

Pupil dilated   Pupil constricted

Fig 2.20 *Left: the response of the eye to dim light. Right: the response of the eye to bright light*

🌳 Biology in context

Some reflex actions are only shown by infants. These actions gradually decrease and disappear after a few months. The 'Moro reflex' occurs when a baby feels as if it is falling. This reflex action causes the arms to spread out and then move back in towards the body.

All newborn mammals show a suckling reflex. They will suck anything that touches the roof of their mouth. These reflexes help to protect infants from harm. For example the Moro reflex may have evolved to help infants maintain a grasp on their mother.

Fig 2.21 *Babies show a palmar grasp reflex, which causes them to grasp anything that touches their palm*

GO! Activities

Activity 2.3.7 Working individually

1. Give the meaning of the term 'reflex action'.
2. Describe the importance of reflex actions.
3. Describe the pupillary light reflex and its importance.
4. Describe the mammalian diving reflex and explain how it benefits these animals.

Activity 2.3.8 Working in pairs

1. The patellar reflex action involves striking the leg just below the knee cap. This causes the lower leg to kick upwards.

   - Work with a partner.
   - Ask your partner to sit with one leg on top of the other and relax.
   - Using the edge of your hand, strike your partner's knee gently, just below the knee cap and watch for a response.

Fig 2.22 *Doctors bring about the knee-jerk reaction using a reflex hammer*

2. Ask your partner to close their eyes for 30 seconds. After 30 seconds tell them to open their eyes. Look for the pupillary reflex response.

I can:

- Explain that reflex actions are fast automatic responses which require no input from the brain.

- Describe examples of reflex actions, for example the pupil of the eye becoming smaller in bright light to protect the sensitive cells at the back of the eye from damage.

- Explain that reflex actions are important to protect the body from harm.

Reflex arc

Reflex actions do not require conscious thought from the brain. However, in most cases, impulses sent to the brain give awareness that the reflex is happening. During a reflex action, **electrical impulses** are sent from a receptor to an effector through a series of neurons. This pathway of neurons is called a **reflex arc**.

If you accidentally touch a hot object, this stimulus is detected by **receptors** in the skin. This causes an electrical impulse to be sent along a **sensory neuron**. The electrical impulse is passed on to a **relay neuron** and then to a **motor neuron**. The motor neuron is connected to an **effector**, in this case a muscle. The impulse causes the muscle to **contract** and moves your hand away from damaging heat. This process is summarised below:

stimulus → receptor → sensory neuron → relay neuron → motor neuron → effector → response

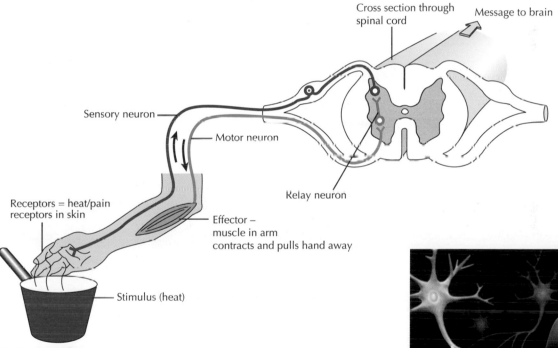

Cross section through spinal cord

Message to brain

Sensory neuron

Motor neuron

Relay neuron

Receptors = heat/pain receptors in skin

Effector – muscle in arm contracts and pulls hand away

Stimulus (heat)

Fig 2.23 *A reflex arc*

In a reflex arc, electrical impulses move from one neuron to another. However, neurons are not in direct contact with each other. Between two neurons there is a small gap called a **synapse.** The electrical impulse from one neuron is passed to the next using **chemicals**. These chemicals (known as neurotransmitters) are released from one neuron and pass across the synapse. They bind onto receptors and initiate an electrical impulse in the second neuron. This process allows electrical impulses to pass from one neuron to another throughout the nervous system.

Fig 2.24 *Chemicals called neurotransmitters move from one neuron to another across a synapse*

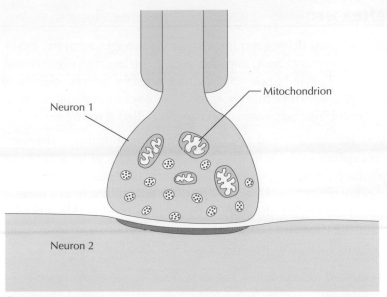

Neuron 1

Mitochondrion

Neuron 2

Fig 2.25 *A synapse*

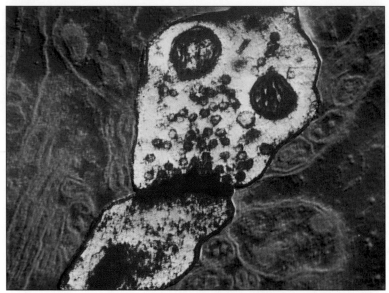

Fig 2.26 *Colour has been added to this electron microscope image. The synapse appears as a dark red line in the centre of the photo*

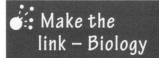

Make the link – Biology

A stimulus is a change in conditions and is detected by receptors (see page **127**).

🔵 Activities

Activity 2.3.9 Working individually

1. Describe the flow of information along a reflex arc.
2. Name the small gaps found between neurons.
3. Explain the role of these small gaps.

Activity 2.3.10 Working in pairs

Make a slideshow presentation to inform someone of your own age about reflex responses and the reflex arc.

Your presentation should include:

- What a reflex action is and some examples.
- An explanation of the terms 'stimulus', 'receptor' and 'effector'.
- Information about a reflex arc including the names of the neurons involved.
- What a synapse is and why it is important.

Activity 2.3.11 Working in pairs

Make a 10-question quiz on reflex actions and the reflex arc. Try to include a mixture of easy and challenging questions. Swap your quiz with a partner and test each other's knowledge.

I can:

- Describe a reflex arc as the passage of an impulse from a sensory neuron, across a relay neuron to a motor neuron.

- State that electrical impulses move along neurons.

- Describe a synapse as a gap between neurons that allows chemicals to transfer from one neuron to another.

Hormones

Multicellular organisms use chemical messengers called **hormones** to send **messages** from one part of the body to another. Hormones are produced by **endocrine glands** and released into the bloodstream. In the bloodstream they can travel to other parts of the body where they have their effect. The tissue a hormone has its effect on is called a **target tissue**.

Target tissues have cells with special receptor proteins on their surface. When a hormone binds onto the receptor protein it brings about changes in the cell. Only tissues with receptor proteins for a specific hormone will be affected by it. One example of a human hormone is anti-diuretic hormone (ADH). This is produced by an endocrine gland called the pituitary gland, found in the brain. ADH travels in the bloodstream to the kidneys. In the kidneys it binds to receptor proteins on the surface of the cells. This alerts them to keep as much water as possible inside the body and only allow a little water to be released with the urine. This hormone helps the body to retain water during times of dehydration.

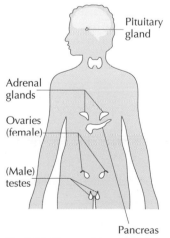

Fig 2.27 *The position of some endocrine glands within the body*

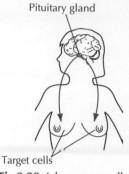

Pituitary gland

Target cells

Fig 2.28 *A hormone called prolactin is produced in the brain and travels to the breasts where it stimulates milk production in women who have just given birth*

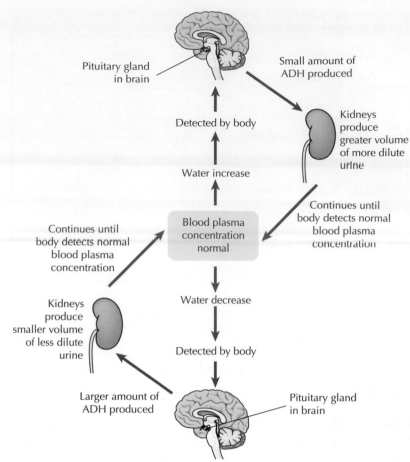

Pituitary gland in brain

Small amount of ADH produced

Kidneys produce greater volume of more dilute urine

Detected by body

Water increase

Continues until body detects normal blood plasma concentration

Blood plasma concentration normal

Continues until body detects normal blood plasma concentration

Water decrease

Kidneys produce smaller volume of less dilute urine

Detected by body

Larger amount of ADH produced

Pituitary gland in brain

Fig 2.29 *The production and role of ADH*

Make the link – Biology

Other molecules such as oxygen and carbon dioxide are transported in the blood (see page **186**).

Biology in context

The hormones oestrogen and testosterone are often thought of as male and female hormones. However, both hormones are found in both sexes. Males produce more testosterone than women, and women produce more oestrogen than men.

In women, oestrogen is mainly produced by the ovaries and has many target tissues including the uterus and breasts. Oestrogen is involved in breast development and regulating the menstrual cycle. In men, oestrogen is required for the testes to produce fully functional sperm.

Men produce testosterone in the testes. It is involved in the growth of the penis during puberty, body and facial hair growth, and the deepening of the voice, and is necessary for normal sperm development. In women, testosterone is produced by the ovaries. It is involved in maintaining muscle mass and bone strength.

GO! Activities

Activity 2.3.12 Working individually

1. Describe the role of hormones in the body.
2. Name the glands that release hormones.
3. Describe how a hormone travels from where it is produced to where it has an effect.
4. Explain why hormones will act only on certain target tissues.

Activity 2.3.13 Working in pairs

Draw or trace an outline of the human body onto an A4 piece of paper.

Research the following hormones:

- ADH
- Adrenaline
- Human growth hormone

On your outline, use a different colour to mark where each of these hormones is made and where it has its effect. Draw a key to show which hormone is represented by each colour.

Activity 2.3.14 Working in groups

The thyroid gland is one of the largest endocrine glands in the body. This gland produces hormones that help to control growth and how the body uses energy. Some people develop an overactive or underactive thyroid gland.

Research each of these conditions. Make an information sheet on an A3 piece of paper to compare and contrast the symptoms associated with these conditions.

✳ Make the link – Health and Wellbeing

Some body builders use testosterone illegally to help increase muscle bulk. This can result in bigger muscles but can also lead to high blood pressure, heart problems or liver damage if used for long periods of time.

I can:

- State that hormones are chemical messengers.
- State that endocrine glands release hormones into the bloodstream where they travel to target tissues.
- State that target tissues have cells with special receptors for the hormone on their surface.
- Explain that only tissues with a corresponding receptor will be affected by a specific hormone.

Control of blood glucose levels

Animals gain most of their energy by respiring **glucose**. This means the cells of the body require a constant supply of glucose to release energy. Glucose is transported around the body in the **blood**. The body uses control mechanisms to keep blood glucose levels constant.

After eating a meal, blood glucose levels increase. This rise in blood glucose levels is detected by the **pancreas**, which increases the release of a hormone called **insulin**. Insulin is released into the blood and travels to the **liver**. Insulin binds to **receptors** on the cells of the liver and activates enzymes which convert **glucose** into **glycogen**. Glycogen acts as a carbohydrate store and this process brings blood glucose levels back down to their normal set point.

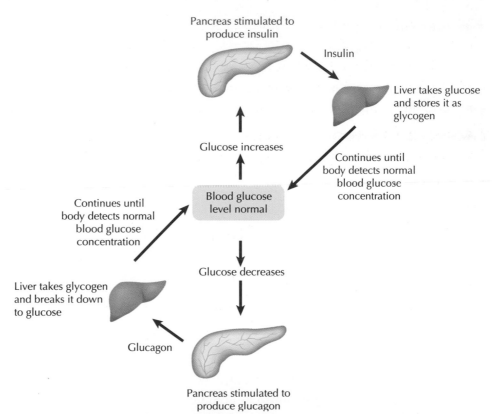

Fig 2.30 *The control of blood glucose levels*

In between meals, blood glucose levels decrease. This change in blood glucose levels is detected by the pancreas. In this case, the pancreas releases a hormone called **glucagon** into the blood. Glucagon travels to the **liver** where it binds to receptors on the cells of the liver and activates enzymes that convert **glycogen** into **glucose**. This raises the blood glucose level back up to its normal set point.

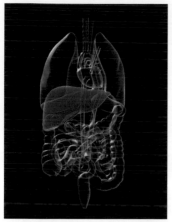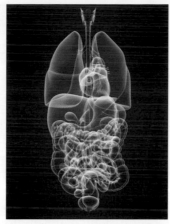

Make the link – Biology

People who suffer from diabetes cannot regulate their blood glucose levels (see page **140**).

All cells require a supply of glucose to carry out respiration (see page **93**).

Fig 2.47 *The liver and pancreas are involved in controlling blood glucose levels*

Activities

Activity 2.3.15 Working individually

1. Name the organ that detects changes in blood glucose levels.
2. Name the carbohydrate that is used to store glucose.
3. Describe the role of insulin in controlling blood glucose levels.
4. Describe the role of glucagon in controlling blood glucose levels.

Activity 2.3.16 Working individually

As well as acting on the liver, insulin has other target tissues. Carry out some research into the role of insulin and write a 100-word report on your findings.

Activity 2.3.17 Working in pairs

With a partner, write down each of the following stages in controlling blood glucose levels onto small pieces of card or sticky notes:

*   Normal blood glucose levels
*   Blood glucose levels increase
*   Increase in blood glucose levels detected by the pancreas
*   Pancreas releases insulin into the blood
*   Insulin travels to the liver
*   Insulin brings about the change of glucose into glycogen
*   Blood glucose levels decrease back to set point
*   Blood glucose levels decrease
*   Decrease in blood glucose levels detected by the pancreas
*   Pancreas releases glucagon into the blood
*   Glucagon travels to the liver
*   Glucagon brings about the change of glycogen into glucose
*   Blood glucose levels increase back to set point

When you are finished, put the cards in the correct order to show the response to a decrease in blood glucose levels. Do the same for an increase in blood glucose.

I can:

- State that a change in blood glucose levels is detected by the pancreas.

- State that glucose can be stored as glycogen in the liver.

- Explain that when blood glucose level increases, the pancreas releases more insulin, which travels to the liver where it activates enzymes in the liver cells to convert glucose into glycogen.

- Explain that when the blood glucose level decreases, the pancreas releases more glucagon, which travels to the liver where it activates enzymes in the liver cells to convert glycogen into glucose.

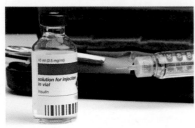

Fig 2.31 *Insulin is made by genetic engineering*

Fig 2.32 *Many diabetics use this pen-like device to deliver insulin rather than a syringe*

Diabetes

Nearly a quarter of a million people in Scotland suffer from **diabetes**. This number has been increasing year on year and is largely due to type 2 diabetes rather than type 1.

People who suffer from **type 1 diabetes** do not produce enough insulin. In some cases they produce no insulin at all. This means they have a high blood glucose level, especially after a meal. Insulin signals body cells to absorb glucose, which can be used to produce energy. In people who suffer from type 1 diabetes this can no longer happen, and the glucose remains in the bloodstream. Type 1 diabetes is treated by injecting insulin regularly throughout the day. Before injecting insulin, a person with type 1 diabetes must check their blood glucose level to ensure they are injecting the correct quantity of insulin.

In people suffering from **type 2 diabetes**, their body cells no longer respond to insulin. This is sometimes known as insulin resistance because the body is capable of producing insulin, however, it does not have any effect on the cells. Like type 1 diabetes, this causes high blood glucose levels. Type 2 diabetes is treated using lifestyle changes. For example eating healthily, exercising regularly and losing weight if necessary. If lifestyle changes alone do not work, medications that lower blood glucose levels may also be given.

It is important that people with both type 1 and type 2 diabetes control their blood glucose levels properly. Left unchecked, high blood glucose levels can cause damage to delicate blood vessels, especially those in the eyes and kidneys. If the blood glucose level remains high over a long period of time this can damage vision or cause kidney failure.

People who have undiagnosed diabetes often experience a symptom called polyuria. This means they urinate more frequently than normal. This symptom is caused by the high levels of glucose in the blood. This high glucose level affects osmosis and draws water into the urine, causing large volumes of urine to be produced. High blood glucose levels can also cause blurry vision. This is caused by high glucose levels in the lens of the eye, which draws water inside.

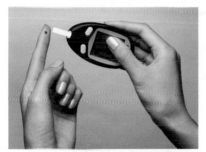

Fig 2.33 *People with type 1 diabetes must check their blood sugar regularly*

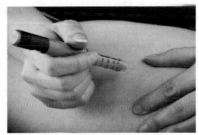

Fig 2.34 *Insulin is injected under the skin (subcutaneously), usually in the leg or stomach*

Biology in context

In some cases, diabetes is diagnosed using a glucose tolerance test (GTT). The patient must not eat or drink for around 12 hours before the test. They are given a measured quantity of glucose and blood samples are taken from them every 30 minutes for 2-3 hours.

A person without diabetes will show a rise in blood glucose level after the glucose is ingested and a quick decrease as the pancreas produces insulin to bring the blood glucose level back down to normal. A person with diabetes will have a high initial blood glucose level which increases further after taking the glucose and remains high for several hours.

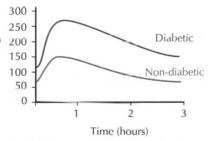

Fig 2.35 *A graph comparing the GTT results of a person with diabetes and a person without diabetes*

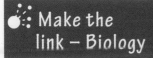

Make the link – Biology

Insulin is made by genetic engineering (see page **72**).

Osmosis is the diffusion of water (see page **30**).

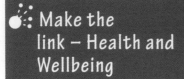

Make the link – Health and Wellbeing

Diabetes is a national problem in Scotland. Type 2 diabetes is often caused by lack of exercise and/or poor diet.

GO! Activities

Activity 2.3.18 Working individually

1. Describe the cause of type 1 diabetes.
2. Describe the cause of type 2 diabetes.
3. Explain the differences between the treatment of type 1 and type 2 diabetes.

Activity 2.3.19 Working in pairs

Frederick Grant Banting and John James Rickard Macleod were awarded the Nobel prize in physiology (medicine) in 1923 for the discovery of insulin. Carry out some research to answer the following questions:

- How was diabetes treated before the discovery of insulin?
- What experiments did the scientists carry out to investigate the function of the pancreas?
- Who was Charles Herbert Best?
- The experiments these scientists carried out involved the use of dogs, but their discoveries have saved countless lives. How do you feel about using animals for experiments like this? Try to think about both sides of the argument.

Activity 2.3.20 Working in pairs

Make an information leaflet or a slideshow presentation that could be given to someone who has just been diagnosed with type 1 or type 2 diabetes. Include information about:

- the cause of their disease
- symptoms of diabetes
- how their disease is treated
- the dangers associated with high blood glucose levels.

I can:

- Describe type 1 diabetes as a disease caused by lack of insulin causing high blood glucose levels, treated by injecting insulin.

- Describe type 2 diabetes as a disease caused by a lack of response to insulin causing high blood glucose levels, treated by diet and exercise.

- Explain that it is important to control blood glucose levels since, if they are high, they can cause damage to blood vessels and affect the functioning of the eyes and kidneys in particular.

- State that uncontrolled blood glucose levels can cause problems with osmosis in cells.

12 Reproduction

You should already know:

- The process of fertilisation involves the male and female sex cells fusing together to form a new organism.
- Cells can have different structures that allow them to carry out different functions.
- Males and females have different reproductive systems.

Learning intentions

- Describe the chromosome complement of gametes.
- Describe the structure of the male and female sex cells and state where they are produced.
- Give details of the male and female reproductive structures, including their function.
- Describe the process of fertilisation in animals.
- Give details of the chromosome complement of a gamete and zygote in animals.
- Name the sex cells found in plants and state where they are produced.
- Describe the reproductive structures found in plants and give their functions.
- Describe the process of fertilisation in plants.
- Give details of the chromosome complement of a gamete and zygote in plants.

Reproduction

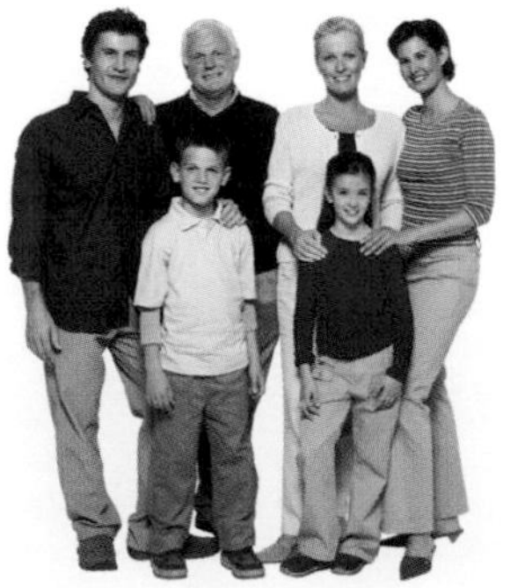

Fig 2.36 *Sexual reproduction produces offspring that are similar but not identical to their parents*

All living organisms must have a method of reproduction to produce new individuals and continue their species. Most animals use **sexual reproduction** to achieve this. Sexual reproduction involves combining genetic information from two individuals (parents) to produce offspring. The offspring produced by sexual reproduction are similar to their parents but not identical.

During sexual reproduction, sex cells from each parent fuse together to form a **zygote**. Sex cells are known as **gametes**. Gametes are **haploid**. This means they have only one set of chromosomes. When the male and female gametes fuse together, the zygote produced is **diploid**. This means it has two sets of chromosomes (fig 2.36).

Sex cells in humans

In animals the male gametes are **sperm cells**. A sperm has a head section containing the nucleus and a tail section that allows it to swim. A sperm cell also has many mitochondria. The female gamete is the **egg**. This is much larger than the sperm cell as it has a large store of food in its cytoplasm (fig 2.37).

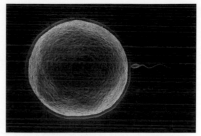

Fig 2.37 *An egg cell is much larger than a sperm cell*

Reproductive structures in humans

The structure of the male reproductive system is shown in Figure 2.38. Sperm are produced in a pair of organs called the **testes**. Sperm travel along the **sperm duct** towards the **penis**. As the sperm move along the sperm duct various glands add fluid to the sperm, forming semen. The penis introduces semen (including sperm) into the female's body to allow fertilisation to take place. Sperm leaves the penis through the urethra (fig 2.38).

The structure of the female reproductive system is shown in Figure 2.39. Eggs are produced in the **ovaries**. They travel along the **oviduct** towards the **uterus**. Sperm are deposited in the

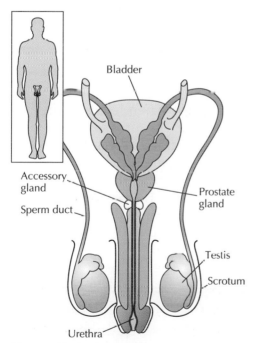

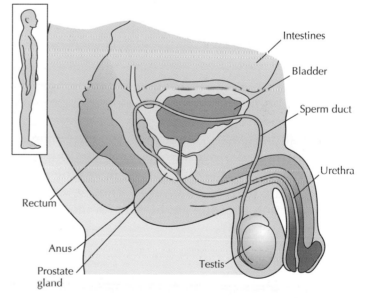

Fig 2.38 *The male reproductive system*

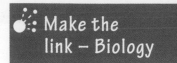

Make the link – Biology

Diploid cells have two sets of chromosomes (see page **39**).

Sperm and egg cells are ideally suited to their functions (see page **114–115**).

vagina. If an egg cell is fertilised by a sperm cell as it travels along the oviduct, it imbeds in the wall of the uterus and develops into a foetus (fig 2.39).

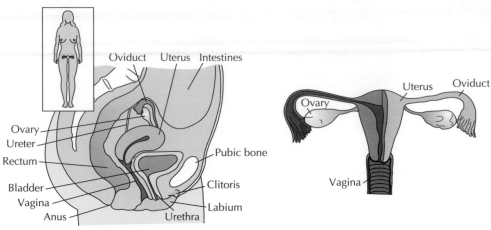

Fig 2.39 *The female reproductive system*

Biology in context

The menstrual cycle is a sequence of changes that take place every month in a woman's body. The menstrual cycle is controlled by hormones. When levels of the hormone oestrogen increase, an egg is released from an ovary. This process is known as ovulation. Oestrogen also causes the lining of the uterus to thicken. Another hormone called progesterone helps the lining of the uterus to remain thickened. If the egg is not fertilised, it travels down the oviduct to the uterus. The thickened lining of the uterus comes away and leaves the body as a period. The period is made up of the unfertilised egg, blood and the lining of the uterus; it usually lasts around 3–7 days (fig 2.40).

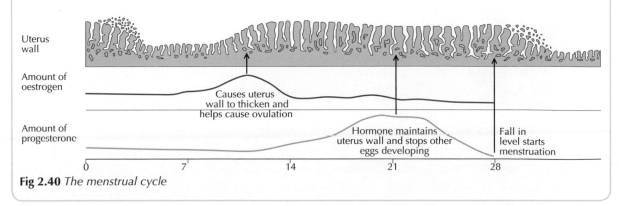

Fig 2.40 *The menstrual cycle*

GO! Activities

Activity 2.4.1 Working individually

1. Asexual reproduction produces offspring that are identical to their parents. Suggest a reason why this might be useful.

2. Give an advantage of sexual reproduction compared to asexual reproduction.

3. Explain why gametes must be haploid.

4. Describe how a sperm cell is suited to its function.

5. Describe how an egg cell is suited to its function.

Activity 2.4.2 Working in pairs

Sketch the male and female reproductive system diagrams onto separate pieces of A5 paper. Label each structure, and below each label, describe its function.

I can:

• Explain that gametes (sex cells) are haploid, meaning they contain only one set of chromosomes.

• State that the male gamete is the sperm, which has a head and tail section and is produced in the testis.

• State that the female gamete is the egg, which is larger than the sperm cell, has a large store of food in its cytoplasm and is produced in the ovaries.

• State that the male reproductive system comprises the testes, which produce sperm, and the sperm duct, which allows sperm to pass to the penis. The penis then releases the sperm from the body.

• State that the female reproductive system comprises ovaries, which produce eggs, and oviducts, which allow the egg to pass to the uterus, where a fertilised egg develops into a foetus. Sperm are deposited in the vagina.

Fertilisation in animals

Fertilisation is the fusion of the male gamete nucleus with the female gamete nucleus. This forms a **zygote**. A zygote is a fertilised egg cell. The male and female gametes are haploid, and the zygote is diploid. Fertilisation in animals can be divided into one of two types: external or internal.

Animals, such as fish, which live in water, often carry out external fertilisation. The female lays unfertilised eggs into the water. The male then releases sperm into the water. The sperm swim through the water and fertilise the eggs. This type of

Fig 2.41 *Like fish, frogs use external fertilisation*

Fig 2.42 *Fertilisation occurs when the male gamete (sperm) nucleus fuses with the female gamete (egg) nucleus*

fertilisation is described as external because it takes place outside the organism's body.

Animals that live on land must use internal fertilisation. This is because sperm move towards egg cells by swimming, and there is no water in a land animal's immediate environment for the sperm to swim in. Sperm are deposited in the vagina along with watery semen. Sperm swim up through the uterus and along the oviducts. Fertilisation takes place in the **oviduct**. The sperm nucleus fuses with the egg nucleus, and a zygote is formed. The zygote begins to divide as it travels along the oviduct to the uterus. It implants in the wall of the uterus and begins to develop (fig 2.43).

Biology in context

After fertilisation the zygote divides many times. When it has formed a ball of 64 cells it is known as an embryo. The embryo implants into the wall of the uterus and begins to develop. The embryo receives nutrients from its mother through the uterus lining. After 3 months it has developed into a foetus and receives nutrients from the placenta through the umbilical cord. In humans, the development of a foetus in the uterus takes 40 weeks. After this time, the foetus is fully developed and ready to be born.

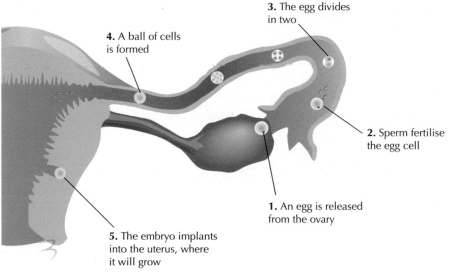

3. The egg divides in two

4. A ball of cells is formed

2. Sperm fertilise the egg cell

1. An egg is released from the ovary

5. The embryo implants into the uterus, where it will grow

Fig 2.43 *Internal fertilisation takes place in the oviduct of the female reproductive system*

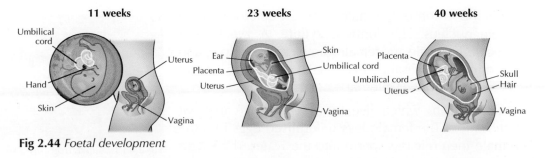

Fig 2.44 *Foetal development*

GO! Activities

Activity 2.4.3 Working individually

1. Starting from where it is produced, describe how a sperm cell fertilises an egg cell.
2. Fish lay hundreds of eggs at a time whereas humans usually release only one at a time. Suggest a reason for this.
3. Which type of fertilisation (internal or external) do you think is most successful? Give reasons for your answer.

Activity 2.4.4 Working in pairs

Pick one of the following stages of human embryo/foetal development:

- Weeks 1 to 10
- Weeks 11 to 20
- Weeks 21 to 30
- Weeks 31 to 40

Make an A4 information sheet including the following:

- A title showing the stage you are investigating.
- Pictures/diagrams.
- Information about what is happening at this stage of development.

I can:

- Describe the process of fertilisation as the fusion of the sperm nucleus with the egg nucleus.
- State that in land animals, fertilisation takes place in the oviduct of the female reproductive system.
- State that in animals such as fish, fertilisation takes place in the water.
- State that gametes are haploid, and when they fuse together during fertilisation, the zygote produced is diploid.

Reproduction in flowering plants

Like animals, plants can use sexual or asexual reproduction to produce new individuals. Many flowering plants use sexual reproduction. Typical flowering plants produce both male and female gametes in the same individual.

The male parts of the flower are the **stamens**. These are made up of an **anther** at the top and a stalk called a **filament**. The anther produces the male gamete, **pollen**. Pollen is usually transferred to other flowers to carry out fertilisation.

The female part of the flower is made up of the **stigma** and the **ovary**. The stigma is where the pollen lands during pollination. The ovary produces the female gametes, the **ovules** (fig 2.48).

As well as male and female structures flowers may also have colourful petals. These attract insects, which can carry pollen from one plant to another. The sepal protects the petals before the flower opens. Flowers may also have nectaries, which produce a sweet liquid called nectar. This encourages insects to visit the flower. In various other species, pollen can be transferred by wind, water and other animals such as birds and bats (fig 2.47).

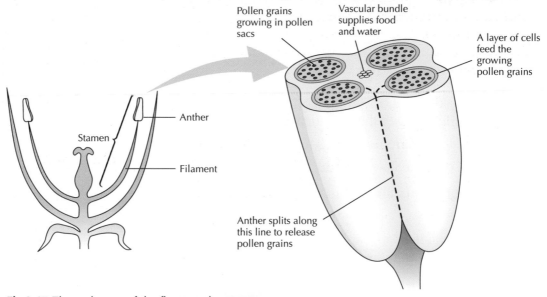

Fig 2.45 *The male part of the flower – the stamen*

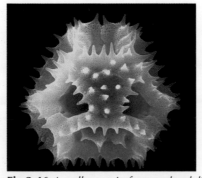

Fig 2.46 *A pollen grain from a dandelion*

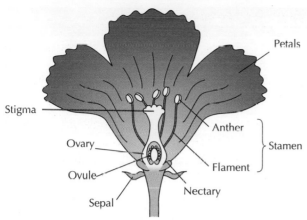

Fig 2.47 *The structure of a typical insect pollinated flower*

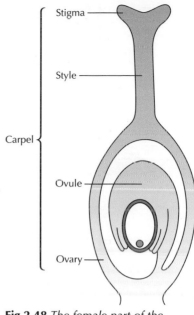

Fig 2.48 *The female part of the flower – the stigma and ovary*

🌳 Biology in context

Some people are allergic to pollen and experience an allergic reaction when it enters their eyes or breathing system. Hay fever is caused by wind-borne pollen. Different people are allergic to different types of pollen and can experience hay fever symptoms at different times of the year. In early spring, pollen from trees is the most common cause of hay fever, while in summer, grasses release pollen, which can affect hay fever sufferers.

Symptoms of hay fever include repeated sneezing, runny/itchy nose, watery eyes, itchy throat and a general feeling of being unwell. Very high pollen counts can also cause breathing to become wheezy. Treatments include antihistamine tablets, which help to reduce the allergic reaction. Anti-inflammatory nasal sprays and decongestant tablets may also be used.

Make the link – Biology

Pollination is the process where pollen is transferred from one flower to another, by the wind or using insects (see page 153)

Make the link – Social Sciences

Many plants rely on insects for pollination and therefore reproduction of their species. Climate change may affect insect populations, which could have knock-on effects on world food production.

Fig 2.49 *A fuchsia flower*

🔵GO! **Activities**

2.4.5 Working individually

Sketch the structure of a flower diagram below onto a piece of A5 paper. Underneath each label, write down the function of the structure.

Activity 2.4.6 Working in pairs

Collect a flower (fuchsia flowers work well) and pull apart the different structures. Try to identify the following parts:

- Stamen (including anther)
- Stigma
- Ovary
- Petals
- Sepal

If you do not have a flower, try to identify these structures using the picture in figure 2.49.

Activity 2.4.7 Working in pairs

The shape, size and surface marking of a grain of pollen are unique to each species of plant. Pollen grains have a tough outer coat, which enables them to survive harsh conditions. Fossilised pollen has been found which was made millions of years ago!

Carry out some research to try and answer the following questions:

- What information does fossilised pollen provide us with?
- The pollen found in honey can be analysed; what will this tell us about the bees that made the honey?
- How can pollen be used to help solve crimes?

I can:

- State that the male gamete in plants is pollen, which is produced by the anthers.

- State that the female gamete in plants is the ovule, which is produced by the ovary.

- State that a flower also has a stigma where the pollen lands, along with petals and a nectary, both of which can be used to attract insects.

Pollination

Before fertilisation can take place in plants, pollination must occur. Pollination is the transfer of pollen from the anther to the stigma. There are two types of pollination. Self-pollination involves pollen moving from the anther to the stigma of the same plant. Cross-pollination involves the transfer of pollen between different plants.

Cross-pollination can be brought about by different methods. Some plants use the wind to transfer their pollen to other plants. The anthers of wind-pollinated plants hang outside the flower. Their petals are small and dull because they do not need to attract insects. Their stigmas also hang outside the flower to catch pollen passing in the wind. Instead of using the wind to carry out pollination, many plants use insects. Plants that use insect pollination usually have brightly coloured petals and produce nectar and scents to attract insects. When insects land on the flower, the pollen becomes attached to them, and they carry it to another plant where it can be brushed off.

Fig 2.50 *An insect-pollinated flower*

Fig 2.51 *A wind-pollinated flower*

Fertilisation in plants

Before fertilisation can take place, the pollen nucleus must reach the ovule nucleus. This occurs through the formation of a pollen tube.

When a pollen grain lands on the stigma, it begins to grow a pollen tube. The pollen tube grows down towards the ovary. The haploid pollen nucleus passes down the tube and fuses with the haploid ovule nucleus forming a diploid zygote. Eventually, the zygote will form a seed, and the wall of the ovary will develop into a fruit (fig 2.52).

Fig 2.52 *A pollen tube extending from a pollen grain that landed on a stigma*

🔵 GO! Activities

Activity 2.4.8 Working individually

1. Is pollen from an insect-pollinated plant likely to cause hay fever? Give a reason for your answer.

2. Give the meaning of the term 'pollination'.

3. Look at the flower in figure 2.53 on the right, and suggest whether this flower is likely to be wind-pollinated or insect-pollinated. Give a reason for your answer.

4. Starting from where it is produced, describe how pollen fertilises an ovule.

Fig 2.53 *A flower*

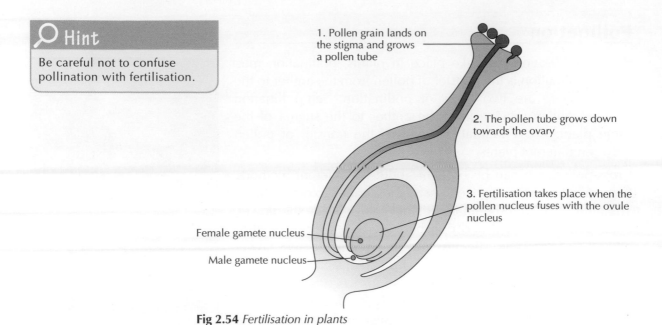

1. Pollen grain lands on the stigma and grows a pollen tube

2. The pollen tube grows down towards the ovary

3. Fertilisation takes place when the pollen nucleus fuses with the ovule nucleus

Female gamete nucleus

Male gamete nucleus

Fig 2.54 *Fertilisation in plants*

Activities

Activity 2.4.9 Working in pairs

Collect an A4 piece of paper and lay it down in a landscape orientation. Draw a line down the middle to create two sections. On one half, write the title 'Fertilisation in animals' on the other half, write the title 'Fertilisation in plants'.

Fill each side with information and diagrams comparing fertilisation in animals with fertilisation in plants.

I can:

- Explain that pollen is transferred from the anther of one plant to the stigma of another in a process called pollination.

- Describe the process of fertilisation in plants as the growth of the pollen tube towards the ovule in the ovary and the fusion of the pollen nucleus with the ovule nucleus.

- State that plant gametes are haploid, and the zygote produced during fertilisation is diploid.

13 Variation and inheritance

You should already know:

- The process of fertilisation involves the male and female sex cells fusing together to form a new organism.
- How to extract DNA from a tissue specimen.
- DNA controls our characteristics and contains the instructions for our cells to work properly.

Learning intentions

- Give some examples of variation within species.
- Describe how sexual reproduction maintains variation.
- Explain the difference between discrete and continuous variation.
- Explain the difference between polygenic and single gene traits and give examples of each.
- Identify examples of dominant and recessive traits.
- Explain what is meant by the term 'phenotype'.
- Give some examples of different phenotypes of a characteristic.
- Explain what is meant by the term genotype.
- Assign a genotype to an individual.
- Explain what is meant by the terms 'homozygous' and 'heterozygous'.
- Use Punnett squares to predict the inheritance of genes.
- Understand the use of pedigree charts to investigate the inheritance of a characteristic.
- Describe the importance of genetic counselling.

Variation

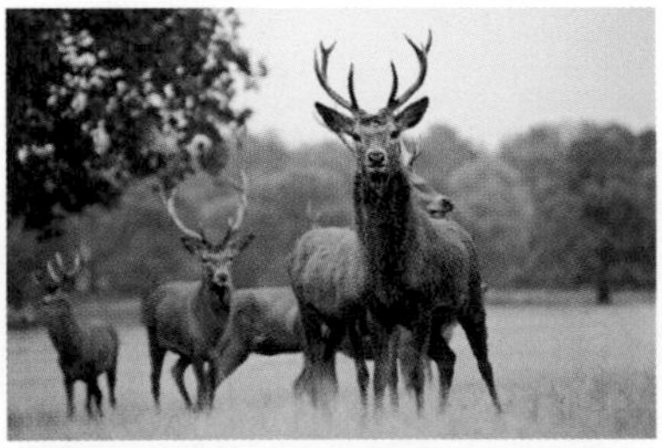

Fig 2.55 *Amongst red deer there are many variations such as antler size or height*

No two people are exactly alike. There are differences between individuals of every species. These differences are known as **variation**. Variation can be seen in animals, for example tail length in red squirrels or coat thickness in highland cattle. Variation also exists in plant species, for example petal colour in primrose plants and leaf size in oak trees.

During fertilisation, genes from each parent are passed onto their offspring. This process, known as **sexual reproduction**, creates an organism that is different from all the other members of its species. This means sexual reproduction contributes to variation within a species.

There are two different types of variation:

- **Discrete variation** is where a characteristic falls into distinct groups, for example the ABO human blood group system. Every human can be placed into one of the blood groups, A, B, AB or O, not in between.

- **Continuous variation** is where a characteristic can have any value in a range, for example height in humans. Humans can range in height from around 60 cm to 250 cm once fully grown.

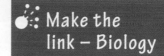

> **Make the link – Biology**
>
> Fertilisation is the process where gametes fuse, forming a zygote (see page 147).

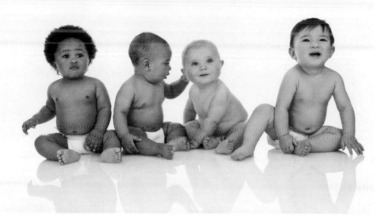

Fig 2.56 *Skin colour varies greatly amongst the human population*

> ◉ **Investigation**
>
> You can see variation in action in your own class. Carry out a survey to find out how many people in your class can roll their tongue. Copy and complete the table below.
>
> | | Can roll tongue | Cannot roll tongue |
> |---|---|---|
> | Number of pupils | | |
>
> Draw a bar chart on a piece of squared paper to display your results.

> **Make the link – Numeracy**
>
> Graph drawing is a skill used in maths as well as biology. You will use different types of graphs in biology, for example bar charts, line graphs and histograms.

GO! Activities

Activity 2.5.1 Working individually

Copy and complete the table below using the following list of characteristics.

- Flower colour in roses
- Shell diameter in limpets
- Height in emperor penguins
- Free or attached ear lobe in humans
- Human hand span
- Gender in house sparrows

Discrete	Continuous

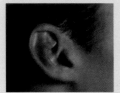

Fig 2.57 *An attached ear lobe*

Fig 2.58 *A free ear lobe*

Activity 2.5.2 Working individually

Bar charts are used to display information about discrete variation. Histograms are used to display information about continuous variation. An example of a histogram is shown below (fig 2.59).

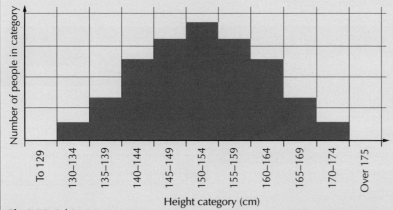

Fig 2.59 *A histogram*

The table below shows information about the hand span of a group of students.

Hand-span category (cm)	Number of students in category
15–16	2
17–18	6
19–20	4
21–22	1

To draw a histogram that displays this information, you will need to use an A5 piece of squared paper.

- Place each hand-span category along a scale on the bottom axis and label it.
- Make a suitable scale for the side axis and label it.
- Plot a bar for each hand-span category. The bars should be touching each other across the chart.

I can:

- Give some examples of variation including flower colour in primrose plants and tail length in red squirrels.

- State that sexual reproduction involves combining genes from two parents.

- State that this process contributes to variation within a species.

- Explain that discrete variation is where a characteristic falls into distinct groups, whereas continuous variation is where a characteristic can have any value in a range.

Polygenic versus single gene traits

Features such as flower colour in scarlet rosemallow and tongue-rolling ability in humans show **discrete variation**. These characteristics are controlled by **single genes**. The ABO blood group system is controlled by one gene. This one gene gives rise to four different blood groups (A, B, AB and O), which show discrete variation.

Most characteristics shown by plants and animals are **polygenic**. This means they are controlled by **many genes** that act together. Polygenic characteristics show continuous variation. Beak depth in finches and height in humans are examples of polygenic characteristics.

Imagine that just three genes control height in humans. Each gene comes in two different forms called alleles. One of the alleles can be given a capital letter symbol and codes for tallness. The other can be given a lower-case letter and codes for shortness. We each have two copies of every gene. Therefore a person who had the alleles AABBCC would be tall, and a person with alleles aabbcc would be small. There would be many people in between these extremes. These people would have different combinations of these genes and therefore different heights (fig 2.60).

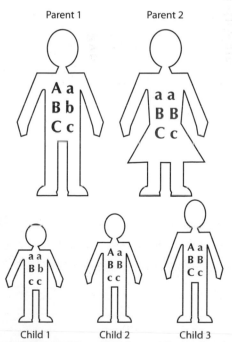

Fig 2.60 *Children often have different heights from their parents. This diagram shows how this is possible*

Biology in context

In some rare cases, non-identical twins have been born with very different skin colours. Polygenic inheritance can explain this situation.

Skin colour is controlled by at least 20 genes. A mixed-race person will have some genes that contribute towards dark skin tone and some that contribute towards light skin tone. A mixed-race parent will pass on a combination of dark and light skin tone genes to their offspring. A twin with dark skin will have inherited more dark skin tone genes, and a twin with light skin will have inherited more light skin tone genes.

Fig 2.61 *Eye colour is controlled by at least three genes. These pictures show blue eyes of varying shades*

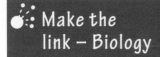

Make the link – Biology

Discrete variation is where a characteristic falls into distinct groups, whereas continuous variation is where a characteristic can have any value in a range (see page 157).

GO! Activities

Activity 2.5.3 Working individually

1. In pea plants, the shape of the peas can be round or wrinkled.
   (a) Name the type of variation shown by this characteristic.
   (b) State whether this is likely to be a single gene or polygenic trait. Explain your answer.

2. In wheat, kernel colour can vary from dark red to white.
   (a) Name the type of variation shown by this characteristic.
   (b) State whether this is likely to be a single gene or polygenic trait. Explain your answer.

Activity 2.5.4 Working individually

Some diseases are caused by alleles of a single gene, whereas others are caused by the alleles of many genes. Carry out some research to discover which of the following diseases are caused by a single gene and which are polygenic.

Copy and complete the table below.

Cystic fibrosis   Asthma

Alzheimer's disease   Sickle-cell anaemia

Hypertension   Schizophrenia

Haemophilia   Diabetes

Single gene	Polygenic

I can:

- State that single gene traits are controlled by one gene and show discrete variation.

- Give examples of single gene traits including flower colour and blood type.

- State that polygenic traits are controlled by many genes and show continuous variation.

- Give examples of polygenic traits including height and eye colour.

Dominant and recessive traits

Genes contain instructions that control the processes taking place inside a cell and the characteristics of the organism. These genes come in different forms called alleles. For example in the pea plant, the gene for flower colour has two different alleles, purple and white.

Diploid organisms have two copies of every gene. Alleles of genes can be described as dominant or recessive.

- Dominant alleles always show up in the appearance of the organism, even if there is only one copy present.

- Recessive alleles only show up in the appearance of an organism if they are paired with another copy of that recessive allele.

In humans, one of the genes that controls hair colour has two alleles, blonde and brown. A person with one brown allele and one blonde allele will have brown hair. This shows that the brown hair allele is dominant, and the blonde hair allele is recessive.

In leopards, spotted coat is dominant to black coat.

> ### Make the link – Biology
>
> Diploid organisms have two matching sets of chromosomes (see page **39**).
> Genes carry instructions for making proteins that control our characteristics (see page **54**).

Two red flower colour genes   One red flower colour gene One white flower colour gene   Two white flower colour genes

Fig 2.62 *Red flower colour is dominant to white flower colour*

Fig 2.63 *Spotted coat is dominant to black coat*

Allele 1	+	Allele 2	= appearance
Spotted coat	+	spotted coat	= spotted coat
Spotted coat	+	black coat	= spotted coat
Black coat	+	black coat	= black coat

Activity

Activity 2.5.5 Working individually

1. Copy and complete the diagram below by filling in the missing word and diagram (fig 2.64).

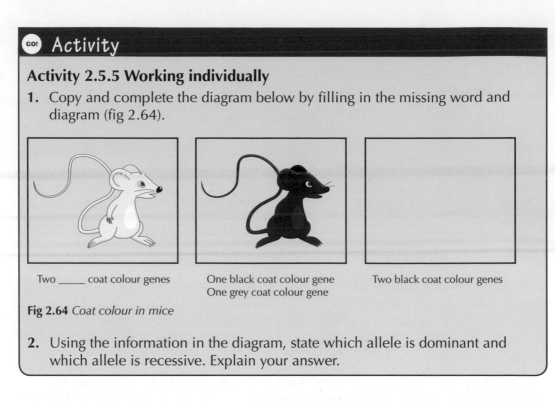

Two _____ coat colour genes

One black coat colour gene
One grey coat colour gene

Two black coat colour genes

Fig 2.64 *Coat colour in mice*

2. Using the information in the diagram, state which allele is dominant and which allele is recessive. Explain your answer.

I can:

- State that dominant genes always show up in the appearance of an organism, even if there is only one copy present in the gene pair.

- State that recessive genes only show up in the appearance of an organism if they are paired with another recessive gene.

- Identify dominant and recessive traits. For example, if a flower has one tall gene and one dwarf gene and appears tall, I understand that the tall gene is dominant, and the dwarf gene is recessive.

Phenotype

Phenotype describes the appearance of an organism. For example in tulips, the phenotype for flower colour may be purple, red, yellow or pink. In humans, the phenotype for ear lobes may be free or attached. Phenotype can sometimes be "hidden", for example human blood groups (fig 2.65).

Fig 2.65 *Tulips have several different phenotypes for flower colour*

Genotype

Certain characteristics are controlled by genetic information passed on from parents to offspring. This genetic information is in the form of genes. Diploid organisms have two copies of every gene, one from each parent. The different forms of a gene are called alleles. For example, the gene that controls blood type in humans has three different alleles, A, B or O. Each person has two copies of the blood-type gene. Together they determine blood group.

An organism's **genotype** describes the set of genes it possesses. When assigning an organism's genotype, we use letters as symbols to represent the different alleles of a gene. Dominant alleles are given capital letters and recessive alleles lower-case letters. The same letter is used for the alleles of one characteristic.

In the fruit fly *Drosophila melanogaster*, one gene that controls body colour has two different alleles – yellow or ebony (black). The allele which results in yellow body colour is dominant to the allele that results in ebony body colour. A fly with two yellow body colour alleles has the genotype YY. A fly with two ebony body colour alleles has the genotype yy. A fly with the genotype Yy has one yellow allele and one ebony allele.

In horses, the agouti gene controls the distribution of black hairs in their coat. The dominant allele (symbol A) causes black hairs to cluster at certain points (i.e. main, tail, lower legs). This produces a bay type horse. The recessive allele (symbol a) causes black hairs to be spread evenly, producing a plain black horse. The table on page 164 shows the different genotypes possible for this characteristic.

Fig 2.66 *A fruit fly showing the dominant characteristic for body colour – yellow*

Fig 2.67 *Horses with a dominant A allele have black hairs clustered at their mane and legs. Horses with two recessive a alleles are black all over*

Genotype	AA	Aa	aa
What this tells us	This horse has two bay alleles	This horse has one bay and one black allele	This horse has two black alleles
Phenotype	Bay	Bay	Black

Biology in context

The fruit fly (*Drosophila melanogaster*) has been used in genetic studies for many years. *Drosophila* is an ideal organism to study genetic crosses as the individuals are small and easy to handle, they have a short generation time, and females produce many offspring at once. *Drosophila* is a model organism. By studying biological processes in *Drosophila*, we may gain some insight into how these processes take place in other organisms, such as ourselves. One very famous scientist to work with *Drosophila* was Thomas Hunt Morgan. Thanks to his work in the early 1900s, he put forward and proved several genetic theories. He was awarded the Nobel prize in physiology (medicine) in 1933 for his research into the role played by chromosomes in inheritance.

Hint

Be careful when deciding on a letter to use when writing out a genotype. It is easy to confuse some capital and lower-case letters such as U and u. Try to use a letter where the capital looks different from the lower case for example G and g.

Fig 2.68 *A 'hitchhiker's' thumb*

Activity

Activity 2.5.6 Working individually

1. (a) Think of an organism (animal or plant).

   (b) Think of a variable characteristic that organism has.

   (c) Draw simple sketches of your organism to show different phenotypes of that characteristic and label each phenotype.

2. One gene controls the ability of the thumb to flex. It has two alleles, 'straight' or 'hitchhiker'. Straight is dominant to hitchhiker. Use the letters H and h as symbols for the possible alleles for this gene.

   Copy and complete the following table.

Genotype	HH		hh
What this tells us		This person has one straight and one hitchhiker thumb allele	
Phenotype			

I can:

- State that phenotype describes the appearance of an organism.

- Give examples of phenotypes of one characteristic. For example, eye colour in humans (different phenotypes may be blue, brown or green).

- State that genotype describes the set of genes an organism possesses.

- Assign a genotype to an organism based on the knowledge that dominant characteristics are given capital letters, and recessive characteristics are given lower-case letters.

Homozygous and heterozygous

The words 'homozygous' and 'heterozygous' are used to describe the set of alleles an organism possesses for one gene – its genotype. If the alleles are the same, we say the organism is **homozygous**. If the alleles are not the same we say the organism is **heterozygous**. For example in pea plants one gene controls pea shape. The allele for round (R) is dominant to the allele for wrinkled (r). A plant with two round alleles (RR) is described as being homozygous. A plant with two wrinkled alleles (rr) is also described as being homozygous. A plant with one round and one wrinkled allele (Rr) is described as being heterozygous (fig 2.69).

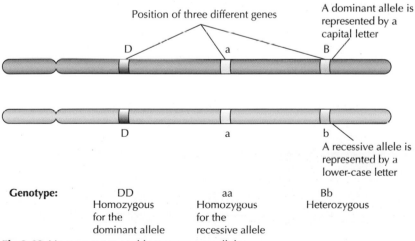

Position of three different genes

A dominant allele is represented by a capital letter

D   a   B

D   a   b

A recessive allele is represented by a lower-case letter

Genotype:

DD	aa	Bb
Homozygous for the dominant allele	Homozygous for the recessive allele	Heterozygous

Fig 2.69 *Homozygous and heterozygous alleles*

Inheritance

We can study the inheritance of a characteristic which is controlled by one gene using a breeding experiment called a monohybrid cross. The original organisms used in a monohybrid cross are called the parental generation (P). The offspring they produce are called the first generation (F_1), and the offspring these individuals produce are called the second generation (F_2).

Gametes are haploid. This means they have only one allele of each gene. In order to determine the genotype of the offspring in a monohybrid cross, we must first be able to determine the alleles that will be present in the gametes produced by the

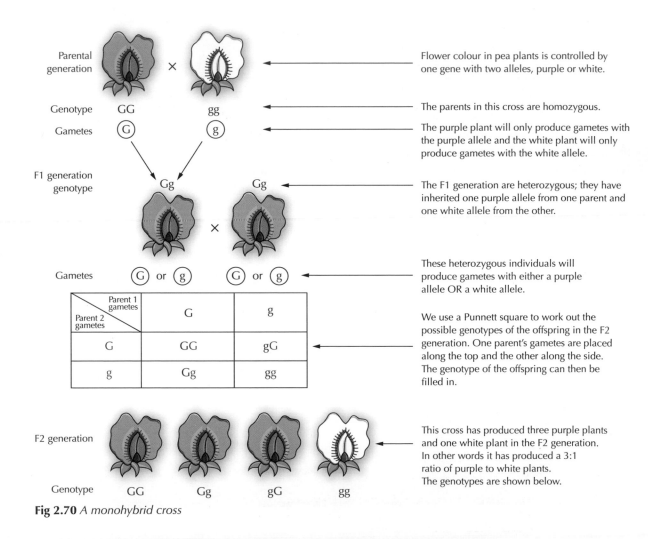

Parental generation		

Flower colour in pea plants is controlled by one gene with two alleles, purple or white.

The parents in this cross are homozygous.

The purple plant will only produce gametes with the purple allele and the white plant will only produce gametes with the white allele.

The F1 generation are heterozygous; they have inherited one purple allele from one parent and one white allele from the other.

These heterozygous individuals will produce gametes with either a purple allele OR a white allele.

We use a Punnett square to work out the possible genotypes of the offspring in the F2 generation. One parent's gametes are placed along the top and the other along the side. The genotype of the offspring can then be filled in.

This cross has produced three purple plants and one white plant in the F2 generation. In other words it has produced a 3:1 ratio of purple to white plants. The genotypes are shown below.

Fig 2.70 *A monohybrid cross*

parents. When the gametes fuse, a new individual is formed with two alleles of the gene that is being studied. One example of a monohybrid cross is shown in figure 2.70.

Make the link – Biology

Gametes (sex cells) are haploid and fuse together during fertilisation (see page 144).

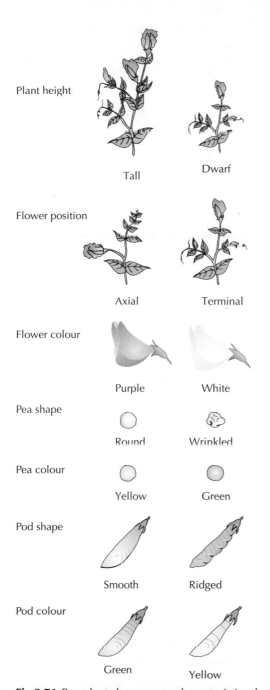

Plant height — Tall / Dwarf

Flower position — Axial / Terminal

Flower colour — Purple / White

Pea shape — Round / Wrinkled

Pea colour — Yellow / Green

Pod shape — Smooth / Ridged

Pod colour — Green / Yellow

Fig 2.71 *Pea plants have many characteristics that can vary between individual plants. Dominant characteristics are shown on the left, and recessive characteristics are shown on the right*

👁 Investigation

One of the features that Gregor Mendel studied in pea plants was pod colour. There are two phenotypes for pod colour: green (dominant) and yellow (recessive).

We can investigate the inheritance of this characteristic using different-coloured beads to represent each allele. Green beads represent the dominant green allele, and yellow beads represent the recessive yellow allele. Follow the instructions below to perform a cross between a homozygous green pod plant and a homozygous yellow pod plant.

- Place 20 green beads into one beaker. This represents the gametes produced by the green pod parent.
- Place 20 yellow beads into another beaker. This represents the gametes produced by the yellow pod parent.
- To model the offspring produced by these parents, take one gamete (bead) from one pot and pair it with one gamete (bead) from the other pot.
- Write down the genotype and phenotype of the offspring, using 'G' for green alleles and 'g' for yellow alleles.

What do you notice about the genotype and phenotype of the offspring in this cross? Can you explain this result?

Planning and designing an investigation

Describe and explain how you could adapt this investigation to show a cross between two heterozygous individuals. (Hint: you will still need 20 beads in each beaker but think about their colour.)

A group of students performed an investigation like this using heterozygous parents. They used the beads to produce 20 offspring.

Making a prediction

What do you think their results would be? How many green pod and how many yellow-pod offspring would you expect there to be?

Results

The results from their investigation are shown below.

Number of green/green pairs in offspring	4
Number of green/yellow pairs in offspring	13
Number of yellow/yellow pairs in offspring	3

Draw a Punnett square for this cross. (Remember the parents were both heterozygous (Gg).)

Conclusion

What phenotype ratio would you expect to see in the offspring?

How do the actual results differ from the expected results?

Evaluation

Explain why the actual results may have differed from the expected results.

How could you adapt the investigation to help improve the results?

🔵 Activities

Activity 2.5.7 Working individually

The positions of three different genes are shown on the diagram below. Copy and complete the diagram by:

- Filling in the genotype for each gene
- Stating whether this individual is homozygous or heterozygous for each gene.

Genotype: _____ _____ _____ _____

Homozygous or heterozygous? _____ _____ _____ _____

Fig 2.72 *Three different genes*

Activity 2.5.8 Working in pairs

Pea shape can differ between individual pea plants. There are two phenotypes for this characteristic, round (dominant) and wrinkled (recessive). Two individual pea plants were crossed. One that was homozygous for round pea shape and one that was homozygous for wrinkled pea shape.

1. Using the symbol 'R' for round and 'r' for wrinkled, write out the genotype of each parent.
2. Write out the gametes each parent could produce underneath.
3. What would be the genotype and phenotype of the offspring (F_1 generation) produced in this cross?

Two of the F_1 generation were then crossed.

1. Write out the genotype of these individuals.
2. Underneath write out the gametes each parent could produce.
3. Draw a Punnett square and fill it in to determine the possible genotypes of the offspring (F_2 generation).
4. How many round pea plants would you expect to see in the F_2 generation?
5. How many yellow pod plants would you expect to see in the F_2 generation?

🌳 Biology in context

Gregor Mendel was an Austrian monk who carried out experiments investigating inheritance in the mid-1800s. Thanks to his work with pea plants, he was able to determine how characteristics are passed on from parents to offspring. Mendel came up with the idea that pea plants had two 'factors' for each characteristic, one inherited from each parent. We now know that these 'factors' are genes. Mendel also noted that some traits that did not show up in an individual pea plant could be passed on to the next generation. In other words he was noticing the effects of dominant and recessive alleles.

Mendel's theories at this time were very advanced, and many scientists misunderstood or disagreed with his ideas. It wasn't until the early 1900s that the importance of Mendel's investigation was recognised. Mendel died in 1884, having never received recognition for his pioneering work (fig 2.73).

Fig 2.73 *Mendel is known as the 'Father of genetics'*

⚛ Make the link – Numeracy

In maths you look at probability and ratios. These are important skills in biology. Punnett squares allow us to work out the genotype and phenotype ratios in the next generation of a genetic cross.

I can:

- Explain that homozygous individuals have two matching alleles of one gene.

- Explain that heterozygous individuals have two different alleles of one gene.

- Work out the genotype of the gametes that will be produced by an individual.

- Use a Punnett square to work out the possible genotypes of the offspring in a monohybrid cross.

Pedigree charts

A **pedigree chart** is a type of family tree diagram. It shows the inheritance of a particular characteristic through several generations in one family. It uses symbols to represent each individual. Males are represented using squares, and females are represented using circles. These symbols are coloured or left blank depending on if the individual shows the characteristic being investigated. A key is always provided to explain the pedigree chart.

One gene that controls earwax type has two alleles, 'wet' or 'dry'. Here is a pedigree chart showing the inheritance of ear-wax type in one family (fig 2.74).

This pedigree chart shows the following information:

- There are three generations in this family study.

- Individuals A and B had three children.

- Couples C and D, and G and H each had two children.

- Individuals F, G, I and M show the dry-earwax phenotype, whereas all the other individuals have the wet-earwax phenotype.

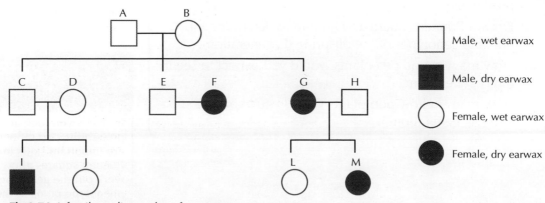

Fig 2.74 *A family pedigree chart for earwax*

- Individuals A and B both have wet earwax but had a child with dry earwax; therefore both of these parents must be heterozygous.

- This also shows that wet earwax is dominant to dry earwax.

Genetic counselling

Some diseases are inherited. If a person has a family history of a genetic disease, they may have concerns about passing the illness on to their children. Doctors can help couples work out the chance of having a child affected by the disease and explain how they can minimise the risk of passing it on. This process is known as **genetic counselling**. Genetic counsellors can also offer advice and support to parents who have a child diagnosed with a genetic condition.

In humans, a pigment called melanin gives the skin, hair and eye irises their colour. In people affected by albinism, there is very little melanin produced, and in some cases none at all. People affected by albinism usually have very pale hair, skin and eyes. As well as the characteristic appearance, their eyes are very sensitive to light, and their eyesight is usually very poor (fig 2.75).

Albinism is an example of a recessive disorder. This means it is possible for two parents to have a normal appearance, but their child could be affected by albinism. In this case the parents are described as carriers. This is because they each "carry" the albinism allele but do not show the disease (fig 2.76).

Fig 2.75 *A Congolese woman and her son who has albinism*

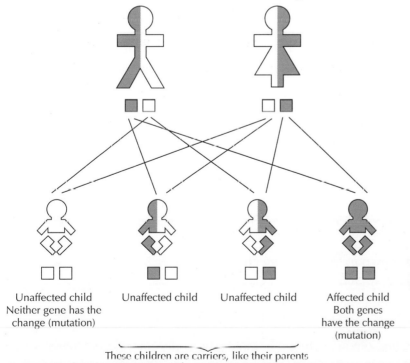

| Unaffected child Neither gene has the change (mutation) | Unaffected child | Unaffected child | Affected child Both genes have the change (mutation) |

These children are carriers, like their parents

Fig 2.76 *A couple who are carriers can have children who are unaffected, a carrier or affected by the disorder*

Make the link – Biology

Dominant characteristics always show up in the phenotype of an organism but recessive characteristics will only show up if there are two recessive alleles present (see page 161).

Biology in context

Queen Victoria reigned over the United Kingdom between 1837 and 1901. She is a direct ancestor of today's royal family (Queen Elizabeth II's great-great grandmother). Queen Victoria was a carrier of haemophilia, a disease that reduces the blood's ability to clot.

Queen Victoria passed the haemophilia allele on to three of her nine children. Two were daughters, who became carriers of the disease and passed the haemophilia allele on to their children. One son received the haemophilia allele and was affected by the disease. He died at age 30 from bleeding after a fall (fig 2.77).

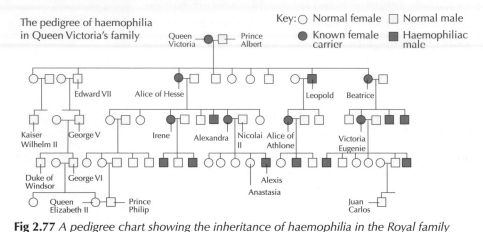

Fig 2.77 *A pedigree chart showing the inheritance of haemophilia in the Royal family*

Activities

Activity 2.5.9 Working individually

Look at the following pedigree chart, which shows the inheritance of freckles. Answer the questions that follow (fig 2.78).

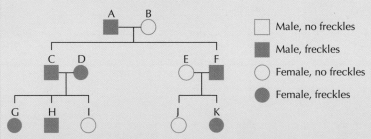

Fig 2.78 *A family pedigree chart for freckles*

1. How many generations are shown in this pedigree chart?
2. How many individuals in this family have no freckles?
3. How many children did couples C and D have?
4. What are the phenotypes of individuals G, H and I?
5. Assuming individual A is heterozygous, which characteristic is dominant: freckles or no freckles?

6. Copy the pedigree chart and assign a genotype to each individual.

7. Are there any individuals for whom you cannot be certain of their genotype? If so why?

Activity 2.5.10 Working in pairs

Cystic fibrosis (CF) is a disease that results in the production of unusually thick mucus, particularly in the lungs. This can harbour infection and cause damage to the delicate tissues of the lungs. (For more information on CF go to page 204.) 1 in 25 people in the UK are carriers of the CF allele.

Imagine you are a genetic councillor. A couple wanting to have a baby have completed a genetic test and are both carriers of the CF allele.

Write a letter to the couple explaining their test results and the implications. Include the following information:

- The results of their test.
- What it means to be a carrier.
- The chance of passing the CF allele on to their offspring.
- Some information about CF in general.
- The options they have if they want to guarantee that their baby will not be affected by CF. (Hint: research pre-implantation genetic diagnosis).

I can:

- Identify individuals in a pedigree chart using the key provided.

- Explain that a carrier is a person who has one copy of a disease allele but is not affected by the disease.

- State that carriers and those at risk of passing on a genetic disorder to their offspring may receive genetic counselling.

14 The need for transport

You should already know:

- Plants use light energy to convert carbon dioxide and water into oxygen and glucose by a process known as photosynthesis.
- Plants play a vital role in sustaining life on Earth.
- The structure and function of key organs and organ systems. For example, the digestive system allows nutrients to be absorbed into the bloodstream and is made up of many organs such as the stomach, liver, pancreas and small intestine.
- The role of technology in monitoring health and improving the quality of life, for example the use of a sphygmomanometer to monitor blood pressure.
- How the body defends itself against disease, for example skin helps to keep microbes out, and acid in your stomach can kill microbes you may eat.

Learning intentions

- Explain the need for transport systems in plants.
- State that water is necessary for transporting materials in plants and photosynthesis.
- Give details of the structure and function of xylem.
- Name the different types of cell found in a leaf and give their functions.
- Describe the structure and function of stomata.
- Explain what is meant by the term 'transpiration'.
- Give details of the flow of water through a plant.
- Describe the structure and function of phloem cells.
- Explain the need for transport systems in animals.
- Give some examples of substances that are transported around the body in the blood.
- Give the function of red blood cells, including the role of haemoglobin.
- Describe how the structure of red blood cells is related to their function.
- Describe the structure and function of arteries, veins and capillaries.
- Identify and name the chambers of the heart.
- Identify and name the major blood vessels leading into and out of the heart.
- State the role of the coronary arteries.
- Describe the pathway of blood through the heart, lungs and body.
- Describe the importance of rings of cartilage in the airways.
- Describe the process of gas exchange and state where it takes place.

- Explain how alveoli allow efficient gas exchange.
- Describe the role of cilia and mucus in protecting the lungs from infection.
- Identify the structures of the digestive system.
- Describe the process of peristalsis.
- Describe the structure of villi.
- Explain the role of villi in the transport of the products of digestion.

Water transport

Water is a vitally important substance for plants. It acts as a raw material for photosynthesis, allowing plants to make their own food. Water also helps plants to absorb minerals from the soil and transport them to where they are needed. Plants must have a **transport system** to allow water to move from where it is absorbed to where it is needed (fig 2.79).

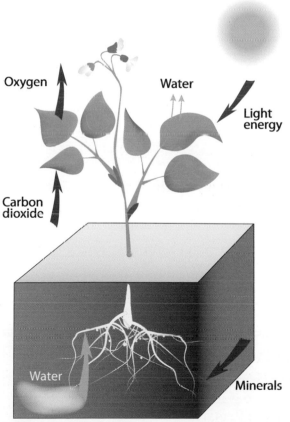

Fig 2.79 *Plants need water to carry out photosynthesis. Water is transported from the roots, where it is absorbed, to the leaves, where photosynthesis takes place*

Xylem

Water moves up through plants in vessels called **xylem**. Xylem vessels are tubes of dead, hollow cells that run through the entire length of a plant. They generally move water in an upward direction, from the roots where it is absorbed to the leaves. Minerals are dissolved in the water and are also transported in xylem vessels. Xylem is often strengthened by a substance called lignin. **Lignin** spans the length of the xylem tubes and takes the form of rings or spirals (fig 2.80).

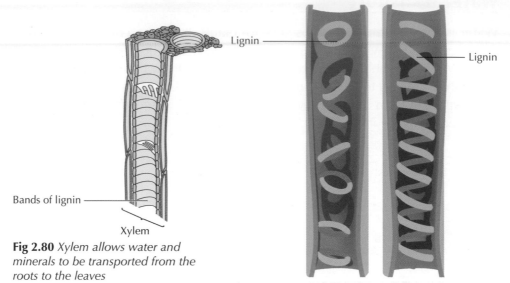

Fig 2.80 *Xylem allows water and minerals to be transported from the roots to the leaves*

Fig 2.81 *Lignin can be found in rings or spirals*

♀ Biology in context

As well as transporting water and minerals, xylem has a secondary function – support. In perennial plants, xylem develops in rings. These rings can clearly be seen in trees. Much of the wood of a tree is made up of xylem tissue. Each year a new ring of xylem develops on the outside of a tree trunk, and this allows us to estimate the age of a tree (fig 2.82).

Fig 2.82 *This image shows the rings on the trunk of a tree, which allow its age to be estimated*

GO! Activities

Activity 2.6.1 Working individually

1. Describe how unicellular organisms gain the materials they need to survive.

2. Organisms that are more than a few cells thick need transport systems. Explain why.

3. Describe the structure of xylem vessels.

4. Describe how the structure of xylem vessels is related to their function.

Activity 2.6.2 Working in pairs

Place a cut stem of celery (with leaves on is best) in a beaker containing dyed water. Leave for 24 hours and then cut it in half across the stem. Have a look at the cut surface that was dipped into the dye (fig 2.83).

(continued)

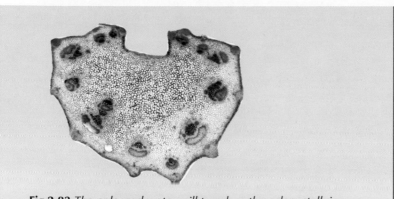

Fig 2.83 *The coloured water will travel up the celery stalk in the xylem vessels*

Make the link – Biology

Photosynthesis requires water and carbon dioxide to produce sugar and oxygen (see page 76).

Hint

Try to remember **x-wam**, xylem transports water and minerals.

I can:

- Explain that plants require transport systems to move substances such as water from one place to another.

- State that plants require water to transport materials and carry out photosynthesis.

- State that water and minerals are transported through plants in xylem vessels.

- State that xylem vessels are strengthened by lignin.

Leaf structure

Leaves are the main site of photosynthesis within a plant. They contain different types of cells to allow them to photosynthesise efficiently.

Section of leaf	Function
Waxy cuticle	A waterproof layer that minimises water loss
Upper epidermis	A layer of colourless cells that allows light to pass through to the next layer
Palisade mesophyll	The main site of photosynthesis; these cells contain lots of chloroplasts
Spongy mesophyll	Cells surrounded by air spaces to allow gases to circulate in the leaf and reach leaf cells
Leaf vein	Contains xylem and phloem, which allow water and sugar to be transported into and out of the leaf
Lower epidermis	Contains guard cells, which form tiny pores called stomata which allow gases to pass in and out the leaf
Guard cells	Controls whether the stomata are open or closed

Fig 2.84 *Leaves are well adapted to carrying out photosynthesis. They have a large surface area to absorb maximum light and are thin to allow light to reach all the cells and gases to exchange readily*

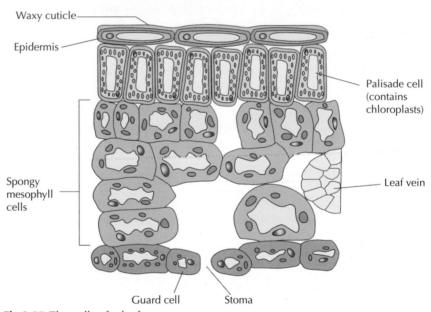

Waxy cuticle

Epidermis

Palisade cell (contains chloroplasts)

Spongy mesophyll cells

Leaf vein

Guard cell

Stoma

Fig 2.85 *The cells of a leaf*

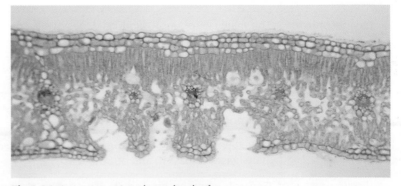

Fig 2.86 *A cross section through a leaf*

Stomata

Stomata (singular: stoma) are pores found on the surfaces of leaves. Stomata allow gases to move into and out of the leaf. Carbon dioxide reaches photosynthesising cells in the leaf by entering through stomata, and oxygen exits in this way. The stomata also allow water vapour to leave the plant by evaporation.

Two guard cells surround each stoma. Guard cells control the opening and closing of stomata. When a plant has little water, the guard cells become plasmolysed, and the stomata close. This conserves water. When a plant has a good supply of water, the guard cells become turgid and the stomata open.

Most of the stomata are found on the lower surface of the leaf. This reduces the volume of water lost by the plant since the stomata are protected from drying winds and sunlight (fig 2.87).

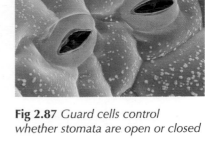

Fig 2.87 *Guard cells control whether stomata are open or closed*

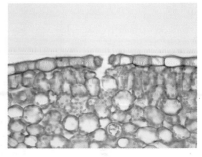

Fig 2.88 *A cross section through a leaf showing one stoma*

GO! Activities

Activity 2.6.3 Working individually

1. Name the layer of colourless cells in a leaf that allows light to pass to the cells below.
2. Explain why spongy mesophyll cells are surrounded by air spaces.
3. Explain how the structure of a palisade mesophyll cell is related to its function.
4. Name the pores found on the surface of a leaf.
5. Describe the role of guard cells with reference to these pores.

Activity 2.6.4 Working individually

Collect a privet leaf and some clear nail polish. Paint a layer of nail polish onto the lower surface of the leaf and allow it to dry. Peel off the nail polish, place it on a slide and look at it under the microscope.

How could this experiment be used to measure the density of stomata on the surface of a leaf?

Activity 2.6.5 Working in pairs

Discuss the following questions with a partner and note down your ideas.

1. Water lilies have leaves that float on the surface of the water in which they grow. Where do you think their stomata are found? Explain your answer.

(continued)

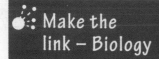

Make the link – Biology

When a leaf cell gains water, it becomes turgid. When it loses water, it becomes plasmolysed (see page 31).

2. Some plants have vertical leaves that are equally exposed to the sun. How do you think this would affect the number of stomata on their upper and lower surfaces?

3. Some plants have very few, small stomata, which reduces the volume of water they lose. What type of habitat might these plants live in?

I can:

- Identify epidermal cells and state that they allow light to pass through to the mesophyll cells.

- Identify mesophyll cells and state that they are the main site of photosynthesis within a leaf.

- State that stomata are pores on the leaf epidermis that allow gas exchange.

- State that the opening and closing of stomata is controlled by guard cells.

Transpiration

Water molecules are small and move from cell to cell by osmosis. In a leaf, water evaporates from the xylem into the air spaces between the spongy mesophyll cells, creating water vapour. If the stomata are open, the water vapour will diffuse out of the leaf into the air. This water loss is known as **transpiration**.

The stomata must be open to allow gas exchange. However, this causes a high level of transpiration. Plants must perform a balancing act to obtain gases for photosynthesis and respiration but ensure they do not lose too much water. One way to achieve this balance is to close the stomata during the night to save water when the plant cannot photosynthesise (fig 2.89).

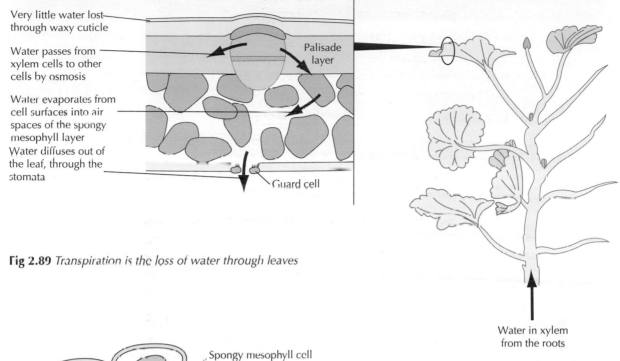

Very little water lost through waxy cuticle

Water passes from xylem cells to other cells by osmosis

Water evaporates from cell surfaces into air spaces of the spongy mesophyll layer

Water diffuses out of the leaf, through the stomata

Palisade layer

Guard cell

Water in xylem from the roots

Fig 2.89 *Transpiration is the loss of water through leaves*

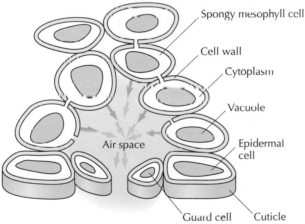

Spongy mesophyll cell

Cell wall

Cytoplasm

Vacuole

Air space

Epidermal cell

Guard cell   Cuticle

Fig 2.90 *Water evaporates into the air spaces in the leaves and moves out through the stomata*

Make the link – Biology

Osmosis is the movement of water molecules from an area of high water concentration to an area of low water concentration through a selectively permeable membrane (see page 30).

◉ Investigation

Fig 2.91 *Transpiration experiment*

The experiment shown can be set up to investigate the effect of different factors on the rate of transpiration. The equipment is weighed, left for 24 hours and then re-weighed. The more water is lost, the higher the rate of transpiration (fig 2.91).

Planning and designing an investigation

A layer of oil is placed over the water surface. Why is this necessary?

Design an experiment that would allow you to investigate the effect of light intensity on the rate of transpiration.

What factors would need to be kept the same in this investigation?

Making a prediction

How do you think light intensity will affect transpiration? Write down your hypothesis.

Results

Here are some sample results from an experiment to investigate the effect of temperature on the rate of transpiration.

Temperature (°C)	Water loss (g)
5	0.45
20	2.25
30	3.30
40	5.40

On a piece of A5 graph paper, draw a line graph to display these results.

Copy and complete the following table to show the rate of transpiration in grams of water lost per hour (remember the experiment was set up for 24 hours).

Temperature (°C)	Rate of transpiration (g per hour)
5	
20	
30	
40	

Conclusion

What do these results show?

What is the relationship between temperature and the rate of transpiration?

Can you explain these results?

Evaluation

How could the reliability of this experiment be improved?

I can:

- Describe transpiration as water loss through the leaves of a plant.
- State that water is lost by evaporation through the stomata.

Activity

Activity 2.6.6 Working individually

Look at Figures 2.90 and 2.91 and write a short note about transpiration. Include the following words in your explanation; evaporation, osmosis, diffusion and transpiration.

Transport of water through plants

Water enters plants by osmosis through root hair cells. It then moves from cell to cell (also by osmosis) until it reaches a xylem vessel. Xylem vessels transport the water up the plant to the leaves. Here it moves from the xylem vessels into the mesophyll cells of the leaf by osmosis. The water then evaporates out of the mesophyll cells into the air spaces in the leaf. Finally it leaves the cell by diffusing out through the stomata (fig 2.92).

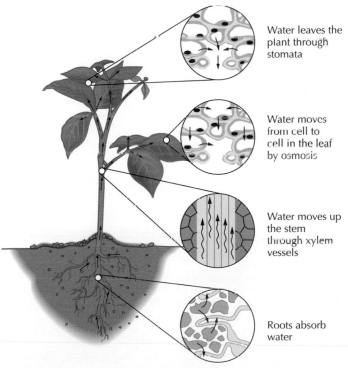

Water leaves the plant through stomata

Water moves from cell to cell in the leaf by osmosis

Water moves up the stem through xylem vessels

Roots absorb water

Fig 2.92 Water moves from the roots up through the plant to the leaves

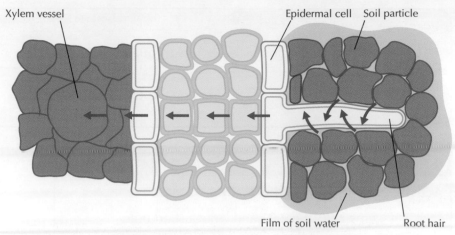

Fig 2.93 *Root hair cells have extensions that project into the soil. This increases the surface area for the absorption*

Transport of sugar through plants

Sugar is transported around the plant in **phloem cells**. Phloem cells are linked together to form a continuous tissue that moves sugar to wherever it is needed. Unlike xylem, phloem cells are alive. Phloem cells are connected by sieve plates that allow substances to pass easily from cell to cell (fig 2.94).

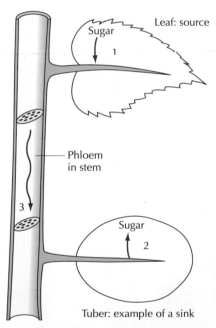

Fig 2.94 *Plants use phloem cells to transport sugar from one part of the plant to another. For example when forming a tuber (e.g. a potato) sugar is transported from the leaves to the growing tuber*

Hint

Remember **PS**: **p**hloem transports **s**ugar.

Fig 2.95 *Phloem tissue in the stem of a plant*

GO! Activities

Activity 2.6.7 Working individually

Copy and complete the table below comparing xylem and phloem.

	Xylem	Phloem
Material transported	Transports water and minerals	
Composition		Made up of living cells
Direction of transport	Transports water and minerals upwards only	
Special features	Contains lignin	

Activity 2.6.8 Working in pars

Make an A4 information sheet about transport systems found in plants. Include information on:

- Why plants need transport systems.
- Structures involved in water and mineral movement from root hairs, to xylem, mesophyll cells and stomata.
- Transpiration.
- Transport of sugar through the plant.

I can:

- Describe the flow of water from the root hair cells, up xylem vessels and into the mesophyll cells of the leaf, finally leaving the plant by diffusing through the stomata.

- State that sugar is transported up and down the plant in living phloem cells.

Transport of substances in the blood

The average adult has 5 litres of blood circulating around their body in blood vessels. The human body takes in useful substances which need to be transported to other parts of the body. It also produces waste, which must be removed. Blood allows substances to be transported from one place to another. The table shows some substances that are carried in the blood.

Substance	Role within the body	Carried from	Carried to
Oxygen	Required to release energy from food	Lungs	All cells of the body
Carbon dioxide	Waste product of respiration	All cells of the body	Lungs
Urea	Waste product from the breakdown of proteins	Liver	Kidneys
Food (e.g. glucose)	Required to produce energy	Small intestine	All cells of the body

Blood is made up of red blood cells, white blood cells, platelets and plasma. White blood cells are part of the body's immune system, and platelets are involved in forming clots when a blood vessel is damaged.

Red blood cells and plasma have important roles in transporting substances around the body. Red blood cells are involved in the transport of oxygen and carbon dioxide. Plasma is the pale yellow liquid part of blood. The main component of plasma is water. Substances such as glucose, carbon dioxide and urea dissolve in plasma and are carried around the body in this way (fig 2.98).

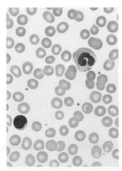

Fig 2.96 *This picture shows a drop of blood under a microscope. Can you see the two white blood cells in amongst the mass of red blood cells?*

Fig 2.97 *Occasionally, patients require plasma transfusions. This picture shows plasma that has been separated from blood cells received from a blood donor*

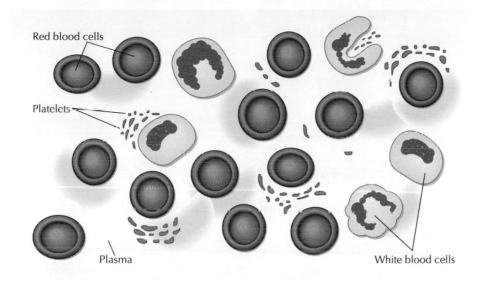

Red blood cells

Platelets

Plasma

White blood cells

Fig 2.98 *Blood is mostly water, with cells and other substances suspended in it*

Biology in context

If a blood vessel in your body becomes damaged, blood can escape the circulatory system. We can see the effects of this and know it as bruising. In order for blood to transport substances around your body, it must stay within blood vessels. If a blood vessel becomes damaged, red blood cells start to leak out. Cell fragments called platelets form a patchwork over the hole in the blood vessel to prevent any more loss of blood. Platelets also bring about the production of a protein called fibrin, which helps to form a clot and stops the bleeding.

A clot, however, is only a temporary plug. Eventually other cells arrive at the wound and carry out a permanent repair. The bruising we see is caused by the red blood cells that have escaped the blood vessel. These red blood cells are removed by white blood cells. During this process they change colour, causing the characteristic colour changes seen as a bruise heals (fig 2.99).

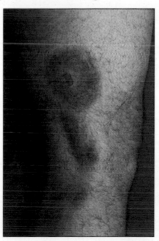

Fig 2.99 *The colour of a bruise is caused by red blood cells that have escaped the circulatory system*

Make the link – Biology

Hormones and antibodies travel from one place to another in the bloodstream (see pages 60 and 135).

Make the link– Health and Wellbeing

Blood transfusions may be given for a number of reasons, for example accidents or disease. Blood transfusions require donations from the public.

GO! Activities

Activity 2.6.9 Working individually

1. Name the two parts of blood that are involved in the transport of substances around the body.
2. a) Give two examples of substances that are transported in the blood.
   b) State where these substances are transported from and to.

Activity 2.6.10 Working individually

Copy the blood drop picture onto a piece of A5 paper. Fill it with information and diagrams about the role of blood and substances that are transported in it (fig 2.100).

Activity 2.6.11 Working in pairs

Search the web to find information and videos about how bruises are made and how the human body heals itself

Fig 2.100 *Blood drop*

I can:

* Explain that blood allows substances to travel around the body.

* State that substances such as oxygen, nutrients and carbon dioxide are transported in the blood.

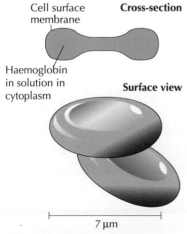

Cell surface membrane

Cross-section

Haemoglobin in solution in cytoplasm

Surface view

7 μm

Fig 2.101 *The special shape of the red blood cell gives it a larger surface area*

Red blood cells

Red blood cells are involved in the transport of oxygen around the body. To help them do this, they contain a protein called **haemoglobin**.

Haemoglobin picks up oxygen in the lungs. When oxygen becomes bound to haemoglobin, a complex called oxy-haemoglobin is formed. Oxygen is released from the oxy-haemoglobin in the tissues to be used for respiration.

Red blood cells are **specialised** to carry out their function. They are very numerous and flexible, and have a special biconcave shape. This shape increases their surface area, allowing diffusion to take place as quickly as possible. Red blood cells do not have a nucleus or mitochondria. This allows the cell to contain the maximum mass of haemoglobin (fig 2.101).

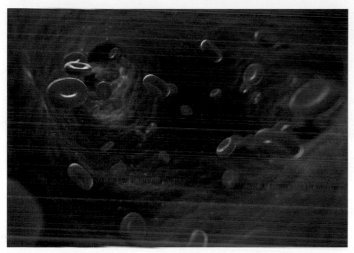

Fig 2.102 *Haemoglobin travels all over the body inside red blood cells*

🔵 Activities

Activity 2.6.12 Working individually

1. Name the protein found in red blood cells that is involved in the transport of oxygen.

2. Copy and complete the equation below:

   h_____ + oxygen ⇌ o_____

3. Describe how a red blood cell is suited to its function.

Activity 2.6.13 Working individually

If a person moves from low altitude to high altitude, the number of red blood cells in their body will increase. At high altitude there is a lower concentration of oxygen in the air, so the body adapts to ensure the cells will receive an adequate oxygen supply. The table below shows the change in number of blood cells in a group of mice after they were moved from 1000 m above sea level to 4000 m.

Time at 4000 m (days)	Average number of red blood cells per mm³ blood (millions)
0	7.5
5	9.5
10	10.5
15	11.0
20	11.0

🌱 Biology in context

Sickle-cell anaemia is a blood disorder that causes red blood cells to have an abnormal shape. The sickle cells can clog blood vessels, leading to pain. The sickle cells also have a shorter lifespan than healthy red blood cells, which can lead to anaemia. Symptoms of anaemia include breathlessness and tiredness, because oxygen supply to the body cells is reduced. There is no cure for sickle-cell anaemia but symptoms can be managed with medications (fig 2.103).

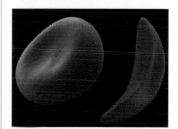

Fig 2.103 *Sickle cells have a very different shape from healthy red blood cells*

(continued)

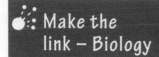

Make the link – Biology

Specialisation is the process whereby a cell becomes suited to its specific function (see page **114**).

Haemoglobin is a protein (see page **118**)

1. On a piece of A5 graph paper, draw a line graph to show the results of the experiment.
2. What has happened to the number of red blood cells in mice transferred to a high altitude?
3. How would this change be an advantage?
4. If a person is moved from low altitude to high altitude quickly, how might their breathing rate change? Explain your answer.
5. Some athletes choose to train at high altitude. How might this give them an advantage during competition?

I can:

- State that red blood cells transport oxygen around the body.
- State that red blood cells contain a substance called haemoglobin that is involved in transporting oxygen.
- Describe how the biconcave shape of a red blood cell increases its surface area so oxygen can diffuse into it as quickly as possible.
- State that red blood cells do not have a nucleus so they can carry as much haemoglobin as possible.

Blood vessels

There are three main types of blood vessels: arteries, veins and capillaries. The structure of each type of blood vessel is suited to its specific function.

Arteries carry blood away from the heart. They have thick, muscular walls to withstand high pressure and a relatively narrow central channel. When the heart pumps blood into an artery, the pressure causes the wall of the artery to push out. When the heart relaxes, the wall rebounds, and this can be felt as your pulse.

Veins return blood to the heart. They have thinner walls and a relatively wider central channel than arteries. The blood in veins is at a much lower pressure than the blood in arteries. Veins contain structures called valves. Blood can flow through valves in one direction only. Valves prevent the backflow of blood.

Capillaries branch from arteries and form networks at tissues and organs. They reunite into veins, which leave the tissues and organs. Capillaries allow substances to be exchanged between the blood and the tissues. To ensure this process happens quickly, the capillary walls are very thin. The walls of capillaries are just one cell thick. Capillaries also have a large surface area. This allows diffusion to take place as quickly as possible (fig 2.104).

Fig 2.105 *Capillaries are so narrow that red blood cells can only just pass through one by one*

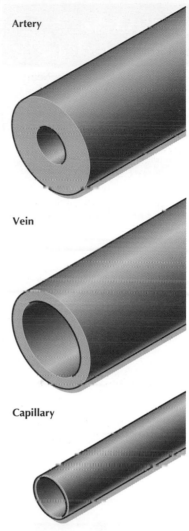

Artery

Vein

Capillary

Fig 2.104 *Arteries, veins and capillaries have very different structures*

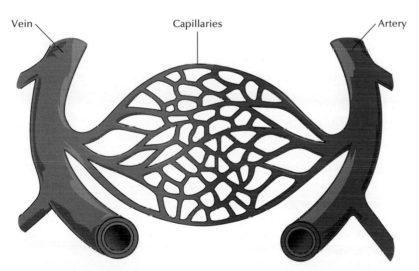

Vein   Capillaries   Artery

Fig 2.106 *Blood leaves the heart in arteries and then flows into capillaries. Materials are exchanged between the blood in the capillaries and the body cells. Blood then flows back to the heart in veins*

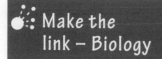

Make the link – Biology

Valves are found in the heart where they prevent the backflow of blood (see page 194).

Hint

Veins carry blood back to the heart and contain **v**alves.

Biology in context

Deep-vein thrombosis is a blood clot that becomes lodged in a vein. This can happen in any vein in the body, most often in deep leg veins. This can cause pain, swelling and redness of the leg. DVT can affect anyone. However, some people are more at risk than others. For example:

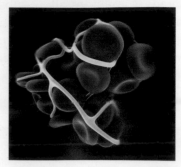

Fig 2.107 *Blood clots are made up of blood cells and other proteins such as fibrin*

- Those who are overweight
- People who have been sedentary for long periods of time, e.g. on a long haul flight
- People with medical conditions such as heart failure.

If a person develops DVT, they may be given anticoagulation drugs, which prevent clots from forming so easily. They may also be advised to raise their leg when resting to prevent blood pooling in the leg. DVT is more likely to happen if blood is circulating slowly, so it is important to take regular exercise to improve circulation. If a person knows they will be sedentary for a long period of time, they should keep hydrated, perform simple leg exercises and take walks whenever possible to reduce the risk of developing DVT (fig 2.107).

GO! Activities

Activity 2.6.14 Working individually

Copy and complete the table below:

Type of blood vessel	Function	Description of vessel wall	Relative width of central channel	Special features
Artery				
Vein				
Capillary				

Activity 2.6.15 Working in groups

Carry out some research into DVT. Produce a short slideshow presentation that could be shown before a long-haul flight. Make sure you explain:

- what DVT is
- how it can be prevented
- the treatments that are used if a person develops DVT

I can:

- Describe the structure of arteries, including thick, muscular walls to withstand blood travelling at high pressure and a narrow, central channel.

- Describe the structure of veins, including thinner walls and a wider, central channel than arteries.

- State that the blood in veins is at a much lower pressure than the blood in arteries, and they contain valves to prevent the backflow of blood.

- State that capillaries form networks at organs and tissues.

- Describe features of capillaries such as thin walls and large surface area to allow the exchange of materials between the blood and body cells.

Structure of the heart

The heart has four **chambers**. The two upper chambers are called **atria** (singular: atrium) and the lower chambers are called **ventricles**.

The heart pumps blood to the lungs, where it picks up oxygen. The oxygen-rich blood travels back to the heart and is then pumped around the rest of the body. The left side of the heart contains oxygenated blood (oxygen-rich) and the right side contains deoxygenated blood (lacking oxygen).

There are four major blood vessels leading into and out of the heart:

- **Vena cava** – veins that carry deoxygenated blood from the body to the right atrium

- **Pulmonary arteries** – arteries that carry deoxygenated blood from the right ventricle to the lungs.

- **Pulmonary veins** – veins that carry oxygenated blood from the lungs to the left atrium.

- **Aorta** – an artery that carries oxygenated blood from the left ventricle to the body (fig 2.109).

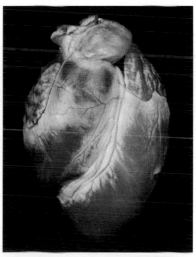

Fig 2.108 *Your heart is about the size of your fist*

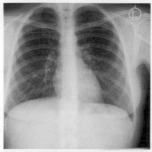

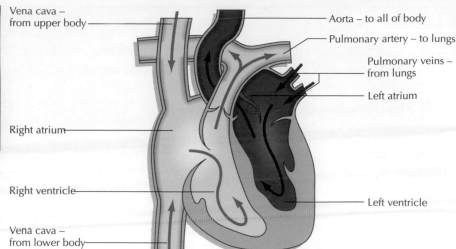

Vena cava – from upper body

Aorta – to all of body

Pulmonary artery – to lungs

Pulmonary veins – from lungs

Left atrium

Right atrium

Right ventricle

Left ventricle

Vena cava – from lower body

Fig 2.110 *The heart is in the centre of the chest*

Fig 2.109 *This diagram shows a vertical section through the heart. Oxygenated blood is shown in red, and deoxygenated blood is shown in blue*

Fig 2.111 *A section through a pig's heart. The string-like structures are tendons, which connect the valves to the heart muscle*

It is important that blood flows through the heart in one direction only. To ensure this happens, the heart has **valves**. Valves are found between the atria and ventricles to ensure the blood flows from each atrium to the ventricle below. Valves are also found at the points where the pulmonary artery and aorta leave the heart. Valves prevent the backflow of blood (fig 2.112).

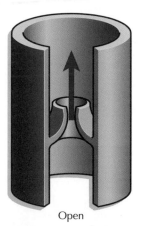 

Open

Closed

Fig 2.112 *Valves snap shut after blood has been forced through them. When you listen to your heart with a stethoscope, the sounds you hear are the heart valves closing*

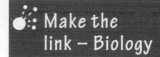 Biology in context

Heart rate is the number of times the heart beats in one minute. The average heart rate at rest is usually between 60 and 80 beats per minute, and this increases dramatically during exercise (see graph below). When the heart rate increases, blood is pumped around the body more quickly. This means the hard-working muscles of the body are supplied with more blood containing glucose and oxygen so they can produce more energy by respiration. The increase in heart rate also allows the blood to carry waste products away from the muscles more quickly (fig 2.113).

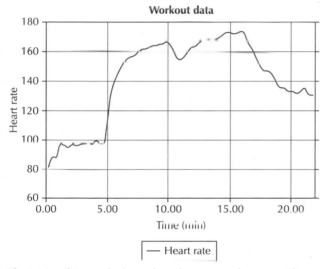

Workout data

Fig 2.113 *This graph shows how heart rate changes with exercise*

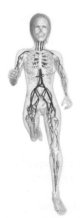

Fig 2.114 *Your cardiovascular system must work much harder when your body is exercising*

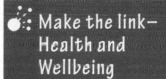 Make the link – Biology

Arteries and veins have very different structures (see page 191).

Make the link– Health and Wellbeing

Your heart rate increases during exercise to pump more blood to your body. Try measuring your heart rate in physical education.

🔍 Hint

Remember as you look at a diagram of the heart, everything is reversed, so left is right, and right is left! Try looking in the mirror and lifting your right hand – it seems to be on the left of your image!

GO! Activities

Activity 2.6.16 Working individually

1. Name the upper chambers of the heart.
2. Name the lower chambers of the heart.
3. Copy and complete the sentence below, selecting the correct option from each bracket.

Oxygenated blood is found in the (left/right) side of the heart, and deoxygenated blood is found in the (left/right) side of the heart.

4. Name the artery that carries blood to the body.
5. Name the veins that return blood from the lungs.

Activity 2.6.17 Working in pairs

Work in pairs through this activity. Type the following address into your web browser:

www.bhf.org.uk/heart-health/how-your-heart-works/know-your-heart.aspx

Click 'start', and on the next screen click 'start' on the 'here to explore' side.

Select 'The heart and what it does' section and work through the topic.

Activity 2.6.18 Working in pairs

How to keep your heart healthy.
1. Maintain a healthy diet and weight
2. Do not smoke
3. Get plenty of exercise
4. Avoid stress

Carry out some research and write a couple of sentences to explain how each of the factors above helps to keep your heart healthy.

Activity 2.6.19 Working in pairs

Heart valves may need to be replaced if they become damaged or diseased. Find out about the two types of replacement valves and write a 200-word report giving details of your findings.

I can:

- Identify the top chambers of the heart as the left and right atria.

- Identify the lower chambers of the heart as the left and right ventricles.

- State that the pulmonary arteries carry blood from the heart to the lungs, and the pulmonary veins carry blood from the lungs back to the heart.

- State that the vena cava carries blood from the body to the heart, and the aorta carries blood from the heart to the body.

Coronary arteries

The heart is a muscle, and all muscles require a blood supply. The heart receives its blood supply from the **coronary arteries**, which are branches of the aorta. The coronary arteries ensure the heart muscle cells have a good supply of glucose and oxygen to carry out respiration and release energy. The heart muscle cells require energy to contract and allow the heart to beat (fig 2.115).

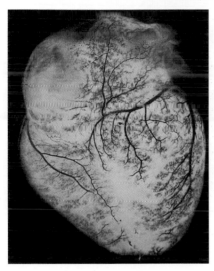

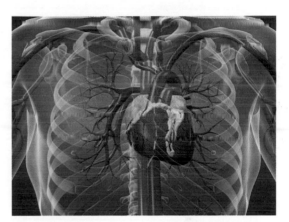

Fig 2.116 *The coronary arteries extend from the aorta and supply the heart muscle with blood*

Fig 2.115 *This picture shows a special X-ray of the coronary arteries. The arteries have been filled with dye so that they can be seen*

Pathway of blood

The circulatory system is made up of the heart and blood vessels. A network of blood vessels extends throughout the body to ensure all cells have an adequate blood supply.

Blood in the **right atrium** is forced into the **right ventricle**. It leaves the heart in the **pulmonary arteries** and travels to the lungs. The blood then returns to the heart in the **pulmonary veins**. The blood enters the **left atrium** and is forced into the **left ventricle**. The blood leaves the heart in the **aorta** and travels to all the cells of the body. Blood returns to the heart in the **vena cava** where it arrives in the **right atrium** (fig 2.117).

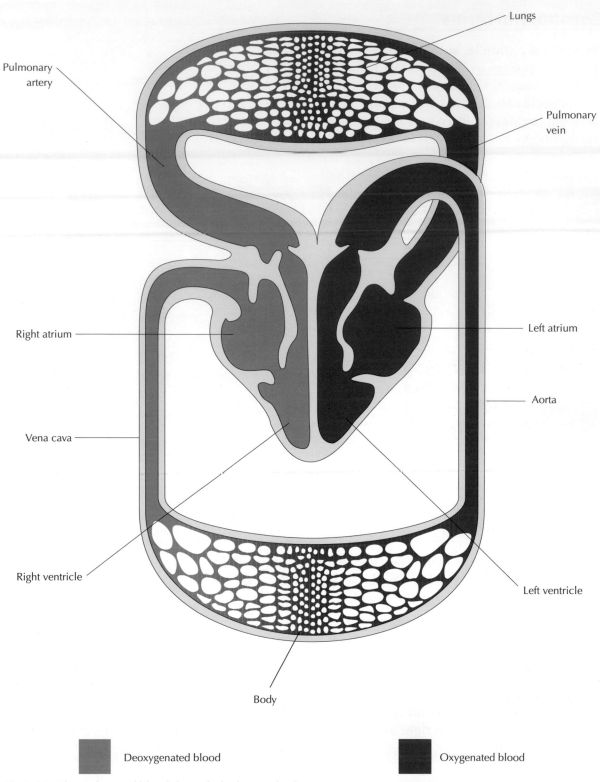

Fig 2.117 *The pathway of blood through the human body*

🌳 Biology in context

When the coronary arteries become partially blocked by fatty deposits, a person may suffer from coronary heart disease (CHD). The fatty deposits cause the walls of the artery to thicken (a process known as atherosclerosis). This may interrupt or block the flow of blood to the heart. If the blood flow to the heart is restricted, this can cause angina (chest pain). If the supply becomes completely blocked, a heart attack may occur.

In 2010 over 8000 people died from CHD in Scotland, and more than 270,000 people were living with CHD. Scotland has the highest rates of CHD in the UK.

CHD cannot be cured but steps can be taken to manage it. These include:

- Lifestyle changes such as keeping physically active and eating healthily.

- Medicines to reduce blood pressure or cholesterol (a fatty substance that contributes to atherosclerosis).

- Surgery to open blocked arteries or bypass graft surgery where an artery or vein from another part of the body (i.e. the leg) is used to bypass the blocked artery (fig 2.120).

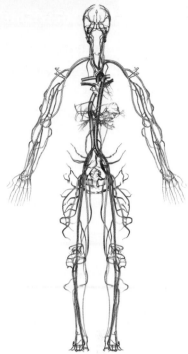

Fig 2.118 *The circulatory system extends throughout the body, ensuring all cells have a good blood supply*

Fatty deposit

Fig 2.119 *The fatty deposits in the arteries reduce blood flow and may cause a clot to form. This can completely block blood flow to the heart and cause a heart attack*

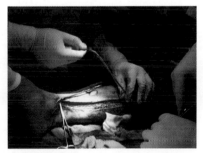

Fig 2.120 *Vessels from other parts of the body can be used to bypass the blocked section of an artery*

⚛ Make the link – Biology

Respiration is the process whereby glucose is broken down and the energy released is used to produce ATP (see pages **90–93**).

🔍 Hint

Remember **a**rteries carry blood **a**way from the heart – Arteries Away.

GO! Activities

Activity 2.6.20 Working individually

1. Give the function of the coronary arteries.
2. Copy and complete the flow chart below:

   right a_____ → right v_____ → p_____ a_____ → lungs → p_____
   v_____ → left a_____ → left v_____ → a_____ → body → v____ c____ →
   rIght a_____.

GO! Activities

Activity 2.6.21 Working in pairs

Work with a partner to carry out some research into CHD.

- Use the Internet to find out information about the causes, symptoms, treatment and prevention of CHD.
- Make a leaflet that could be displayed in a doctor's surgery to inform the public about CHD.
- Try to include as much information as possible and make your leaflet look colourful and eye-catching.

Activity 2.6.22 Working in pairs

Imagine you are a red blood cell in the right atrium. Describe the pathway you would take around the body by:

- writing an imaginative story of 200 words

  or
- making a comic strip with 6 panels

I can:

- State that the coronary arteries supply the heart muscle with blood.

- Describe the flow of blood around the body, right atrium → right ventricle → pulmonary arteries → lungs → pulmonary veins → left atrium → left ventricle → aorta → body → vena cava → right atrium.

Breathing system

When we take a breath in, air moves into our lungs through the breathing system. Air enters the breathing system through the nose and mouth. It then passes down the trachea to the lungs. As the air moves into the lungs, it travels through a series of smaller branched passages: first the bronchi (singular: bronchus) and then the bronchioles. At the end of the bronchioles are air sacs known as alveoli (singular: alveolus). In the alveoli, oxygen moves into the blood, and carbon dioxide moves out. This process is known as **gas exchange**. To allow gas exchange to take place, the alveoli are covered in blood capillaries. To allow gas exchange to take place continuously, the larger air passages are held open by rings of cartilage (fig 2.121).

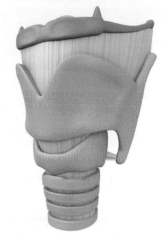

Fig 2.122 *An airway held open by cartilage*

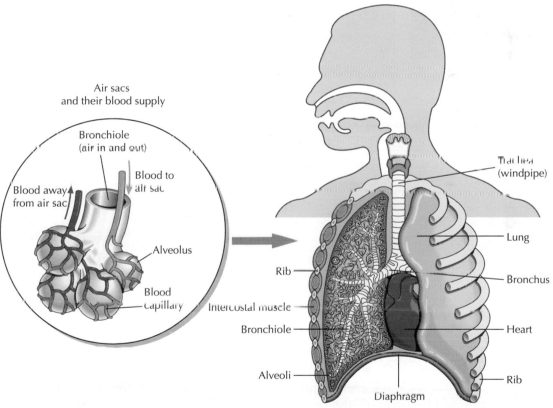

Fig 2.121 *The breathing system*

Alveoli

Alveoli are the site of gas exchange in the lungs. They allow oxygen and carbon dioxide to diffuse into and out of the blood. All the cells of the body require oxygen, and carbon dioxide is a waste product. It is important that gas exchange happens quickly and efficiently. The alveoli are adapted for **efficient gas exchange** in several ways:

- There are millions of alveoli in the lungs, increasing the surface area for diffusion to take place.

- The alveoli are surrounded by a dense network of capillaries, giving them a good blood supply.

- The walls of the alveoli are very thin, meaning diffusion must take place only over a very short distance.

- The lining of the alveoli is moist to allow gases to dissolve before they diffuse (fig 2.124).

Make the link – Biology

Oxygen is required to break down glucose (see page 93).

Oxygen diffuses from a high concentration in the alveoli to a low concentration in the blood (see page 27).

Fig 2.123 *The alveoli create a surface area equal to the size of a tennis court*

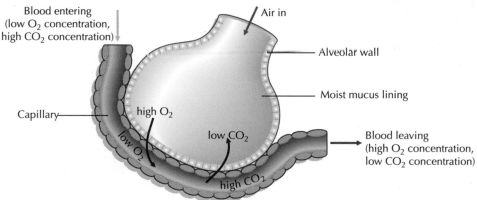

Blood entering (low O_2 concentration, high CO_2 concentration)

Air in

Alveolar wall

Moist mucus lining

Capillary

high O_2

low O_2

low CO_2

high CO_2

Blood leaving (high O_2 concentration, low CO_2 concentration)

Fig 2.124 *Gas exchange takes place in the alveoli*

Biology in context

The ribcage, intercostal muscles and diaphragm are all involved in the mechanism of breathing. When we breathe in, the intercostal muscles contract, causing the ribcage to move up and out. At the same time, a large sheet of muscle below the lungs called the diaphragm contracts and moves down. Together these movements allow air to be drawn into the lungs. When we breathe out, the intercostal muscles relax, causing the ribcage to move down and the diaphragm also to, allowing air to be pushed out of the lungs (fig 2.125).

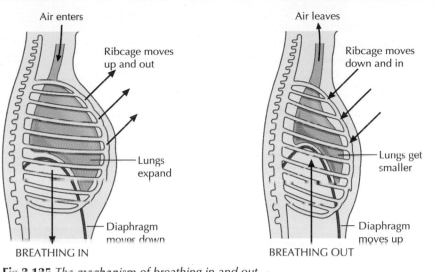

Fig 2.125 *The mechanism of breathing in and out*

GO! Activities

Activity 2.6.23 Working individually

1. Starting with the trachea, name the passages a molecule of oxygen must pass through to reach the blood.

2. Explain the role of cartilage in the breathing system.

3. Copy the following table into your notes and complete it using the information on page 202.

Adaptation of the alveoli	Importance in allowing efficient gas exchange
Millions of alveoli	
Alveoli walls are very thin	
Lining of the alveoli is moist	
Alveoli surrounded by a network of capillaries	

Activity 2.6.24 Working in pairs

Copy the following link into your browser:

www.bbc.co.uk/learningzone/clips/anatomy-and-physiology-of-the-lungs/5373.html

Watch the clip showing a dissection of a pig's lung and write down three interesting facts about the lungs.

Activity 2.6.25 Working individually

Place two fingers on the front of your neck and move them up and down. The bumps you feel are the rings of cartilage which keep your windpipe (trachea) open at all times. Think about the tubes on a vacuum cleaner – are they similar to your windpipe?

I can:

- State that rings of cartilage hold the airways open at all times.

- State that gas exchange in the lungs involves the movement of oxygen into the blood and carbon dioxide out of the blood.

- State that gas exchange takes place in the alveoli.

- Describe features of the alveoli such as large surface area, thin walls and good blood supply, which allow efficient gas exchange.

Mucus and cilia

Fig 2.126 *This photo shows cells that produce mucus (orange) and cilia (green)*

With every breath, we take air into our lungs that contains dust, dirt and microorganisms. The lungs need to have a way of dealing with this situation or they risk becoming infected or damaged.

One method of protecting the lungs is the production of **mucus**. Special cells in the lining of the airways produce sticky mucus, which traps dirt and microorganisms. This prevents them from reaching the delicate tissues of the lungs and causing damage or infection. Mucus is produced continuously.

Other cells in the lining of the airways have tiny hair-like structures called **cilia**. Cilia beat upwards in a wave-like motion. This moves mucus (containing dirt and microorganisms) up and away from the lungs to the nose and throat where most of it is swallowed subconsciously (fig 2.126).

🌳 Biology in context

Cystic fibrosis (CF) is an inherited disorder. People with CF have two faulty copies of a gene, which results in the production of unusually thick mucus. This mucus can block the narrow passages of the lungs and cause infection and damage. Around 1 in 25 people in the UK are carriers of CF. This means they have one copy of the faulty gene but are not affected by the disease. If two carriers have a child, however, both parents may pass on the faulty gene, and their child would suffer from cystic fibrosis. Symptoms of CF include frequent chest infections and difficulty breathing. Cystic fibrosis also affects the pancreas, which can become blocked due to the thick mucus produced by the body. There is currently no cure for cystic fibrosis. However, symptoms can be managed. Patients may be given antibiotics continuously to prevent bacterial infections, and if the lungs begin to fail, a transplant may be necessary. Research is being carried out into ways of curing cystic fibrosis by a process called gene therapy. This means replacing the faulty gene that causes the disease with a working copy (fig 2.127).

Fig 2.127 *People with cystic fibrosis may need to inhale medications to keep their lungs as healthy as possible*

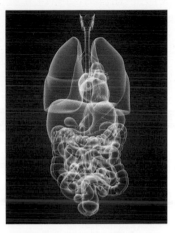

Fig 2.128 *Cystic fibrosis can affect the pancreas (highlighted here in red) as well as the lungs*

Activities

Activity 2.6.26 Working individually

Draw a diagram to show the two types of cell found in the airways of the lungs. Underneath your diagram, explain how the types of cell in the airways help to protect the lungs.

Activity 2.6.27 Working in pairs

If a couple find out they are carriers of CF, they may worry about having a child with the disease, and this could affect their decision to have children. One option open to couples in this situation is to use a process called pre-implantation genetic diagnosis (PGD). PGD allows scientists to screen the couple's embryos and implant the mother with healthy embryos only, ensuring the couple will have a baby who is not affected by CF (fig 2.129).

To carry out PGD, eggs are removed from the mother and fertilised with the father's sperm, creating an embryo. A cell is removed from the embryo and its DNA is screened. If it is normal, the embryo is implanted into the mother's uterus and given the opportunity to develop into a healthy baby.

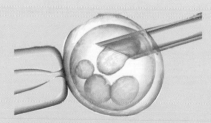

Fig 2.129 *Pre-implantation genetic diagnosis (PGD)*

Make the link – Biology

A carrier is a person who has one copy of the faulty gene but does not have the disease (see page 171)

Hint

Mucus acts like a conveyor belt, which is moved by the beating cilia.

(continued)

1. PGD is a very expensive process. Do you think it should be available for free on the NHS? Explain your answer.

2. During the PGD process, any unhealthy embryos will be destroyed. Do you think this is ethical? Explain your answer.

3. PGD can be used to determine the sex of an embryo. This allows parents to decide the sex of their child. Do you think this should be allowed? Explain your answer.

I can:

- State that mucus in the airways traps dust and microorganisms.
- State that cilia move the mucus up and out the lungs.

The digestive system

The digestive system allows large food molecules to be broken down to allow absorption into the blood stream. Food enters the digestive system through the mouth and travels along the oesophagus to the stomach. Food begins to be broken down by digestive enzymes in the stomach before it moves into the small intestine. In the small intestine, the food is further broken down, and the products are absorbed into the bloodstream. The remains of the food then pass into the large intestine, where water is absorbed, leaving material that becomes faeces. Faeces are stored in the rectum before leaving the body through the anus.

Several organs are involved in the digestive system. The liver produces bile, which helps to break down fats. Bile is stored in the gall bladder before it is released into the small intestine. The pancreas also helps digestion by producing enzymes, which travel to the small intestine and break down large food molecules into smaller ones (fig 2.130).

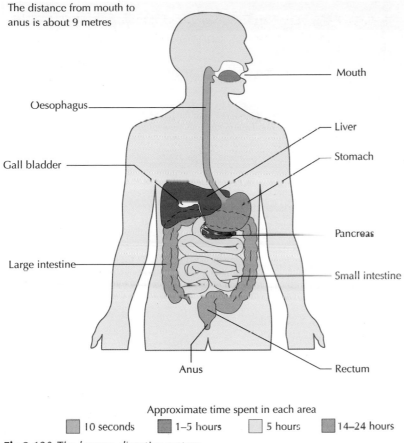

The distance from mouth to anus is about 9 metres

Mouth

Oesophagus

Liver

Stomach

Gall bladder

Pancreas

Large intestine

Small intestine

Anus

Rectum

Approximate time spent in each area

| 10 seconds | 1–5 hours | 5 hours | 14–24 hours |

Fig 2.130 *The human digestive system*

Peristalsis

When you swallow, food does not fall down your digestive system; it is pushed. The tubes of the digestive system are surrounded by muscles that push the food along. This process is called **peristalsis**. Peristalsis allows food to move from the mouth to the anus. The muscles behind the ball of food contract and the muscles in front relax. This allows food to move forward through the digestive system (fig 2.131).

> **Make the link – Biology**
>
> Enzymes are used to speed up chemical reactions. In the digestive system they are used to aid digestion (see page **60**).

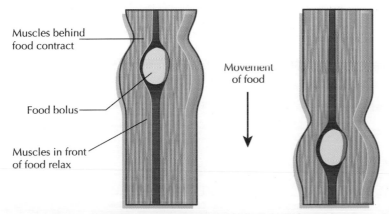

Muscles behind food contract

Movement of food

Food bolus

Muscles in front of food relax

Fig 2.131 *Muscles behind the ball of food contract, and muscles in front relax. This process is called peristalsis*

GO! Activities

Activity 2.6.28 Working individually

1. Describe the path that food takes through the digestive system, from the mouth to the anus.

2. a) Name the process that allows food to move along the digestive system.

   b) Describe this process.

Activity 2.6.29 Working in pairs

Type the following address into your web browser and watch the digestion animation. Click 'zoom in' to see what is happening at each stage:

www.kitses.com/animation/swfs/digestion.swf

Activity 2.6.30 Working in pairs

Have a go at modelling peristalsis. Collect a length of Bunsen burner tubing, some washing-up liquid and a marble. Squeeze a drop or two of washing-up liquid into the tube and then try to push the marble through the tube with your fingers. How did you do it? How is this similar to peristalsis (fig 2.132)?

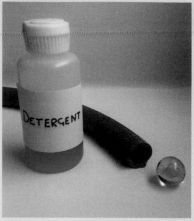

Fig 2.132 *Use this equipment to model peristalsis*

I can:

- State that food is moved through the digestive system by peristalsis.

- Describe the process of peristalsis as contraction of the muscles behind the food and relaxation of the muscles in front.

Villi

Digestion breaks down large insoluble molecules into small soluble molecules that can be absorbed into the bloodstream. The absorption of small molecules occurs in the small intestine. The interior of the small intestine is covered in projections called **villi** (singular: villus). Villi contain a **network of blood capillaries** to absorb glucose and amino acids, and a vessel called the **lacteal** to absorb the products of fat digestion (fatty acids and glycerol).

The small intestine is adapted to make absorption as efficient as possible.

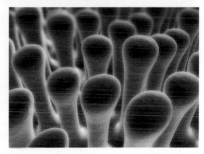

Fig 2.133 There are millions of villi in the small intestine

- There are millions of villi in the small intestine that provide a large surface area for absorption of food molecules.

- Each villus has a very thin wall, which allows molecules to diffuse quickly into the body.

- The villi have a good blood supply and a lacteal, which allows food molecules to move from the small intestine to the rest of the body. This also maintains a concentration gradient between the food molecules in the intestine and the bloodstream to allow diffusion to take place as quickly as possible (fig 2.134).

Contains a network of capillaries

The products of protein digestion (amino acids) and carbohydrate digestion (glucose) are absorbed into the capillaries

Villus wall is only one cell thick

The products of fat digestion (fatty acids and glycerol) are absorbed into the lacteal

Blood arriving at the villus to pick up food molecules

Blood leaving the villus, taking the food molecules to the rest of the body

Fig 2.134 A villus

Fig 2.135 A magnified image of the lining of the small intestine

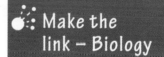

Make the link – Biology

Diffusion is the movement of gas or liquid molecules from an area of high concentration to an area of low concentration (see page 27).

Biology in context

Coeliac disease is an inherited disorder, where sufferers are intolerant to a substance called gluten, which is found in wheat. If a person with coeliac disease eats wheat, this can result in a shortening of their villi. The damage to the villi is caused by the body's own immune system attacking them. This type of condition is known as an autoimmune disease.

Symptoms of coeliac disease include diarrhoea, abdominal pain, weight loss and lack of energy. If a person shows a combination of these symptoms, a blood test or biopsy from their small intestine can

Fig 2.136 *People with coeliac disease must avoid wheat products or risk causing damage to their villi*

confirm if they are suffering from coeliac disease. Currently, the only treatment for this condition is to follow a gluten-free diet, which allows the intestines to heal and eases the symptoms associated with the disease (fig 2.136)

Make the link– Health and Wellbeing

In food and technology, you consider the dietary needs of different individuals. A person with coeliac disease must follow a strict gluten-free diet.

GO! Activities

Activity 2.6.31 Working individually

Copy the following table into your notes and complete it using the information above.

Adaptation of the small intestine lining	Importance in allowing absorption
Millions of villi	
Villi walls are very thin	
Villi have a good blood supply and a lacteal	

Activity 2.6.32 Working in pairs

A person who is unaware they have coeliac disease can cause damage to the villi in their small intestine.

1. One symptom they may show is weight loss. How does this symptom relate to the damage caused by the disease?

2. Carry out some research into anaemia. How might coeliac disease cause anaemia?

I can:

- State that the villi contain a blood capillary to absorb glucose and amino acids, and a lacteal to absorb the products of fat digestion.

- Describe the features that allow villi to carry out absorption efficiently, including having a large surface area and thin walls.

15 Effects of lifestyle choices on animal transport and exchange systems

You should already know:

- The structure and function of key organs and organ systems. For example, the heart is a muscle which pumps blood around your body.
- The role of technology in monitoring health and improving the quality of life, for example the use of a sphygmomanometer to monitor blood pressure.

Learning intentions

- Explain how poor diet and lack of exercise can affect transport and exchange systems causing diseases.
- Explain how smoking tobacco and drinking alcohol can affect transport and exchange systems, and how this can cause diseases.

Fig 2.137 *Foods that are high in fat increase the levels of cholesterol in the blood*

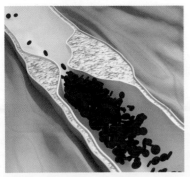

Fig 2.138 *Plaques forming in arteries can severely restrict blood flow*

Diet

We all need a certain quantity of **fat** in our diets. Fats are a good source of energy, they are used to insulate our bodies, and they form part of the structure of cells. Problems arise when we eat too much fat. A high-fat diet makes us more likely to become overweight and can increase the levels of a substance called cholesterol in our blood. Cholesterol can be deposited in the arteries along with other fatty substances, forming plaques. This causes a condition known as atherosclerosis. A patient with atherosclerosis, has narrowed arteries and restricted blood flow to certain organs. If the coronary arteries are affected by atherosclerosis this can be especially dangerous. If a plaque ruptures, a blood clot may form. Blood clots can block the blood supply to the heart muscle and trigger a heart attack (fig 2.137).

It is not only fatty foods that cause problems for the circulatory system. Eating lots of **salty foods** can increase blood pressure. Blood pressure is the force of blood pushing against the walls of arteries. High blood pressure (known medically as hypertension) puts extra strain on the heart and arteries. People with hypertension are at an increased risk of suffering a heart attack or stroke (fig 2.139).

Exercise

It is important to lead an active lifestyle and **exercise regularly**. When we exercise, blood flow to the muscles increases, and frequent exercise will improve circulation. Good circulation protects against conditions such as DVT (for more information on DVT see page **192**).

Obesity is usually caused by excess body fat, and lack of exercise is a major contributory factor. People who are obese are at a much higher risk of developing diseases such as type 2 diabetes, heart disease and colon cancer. By exercising regularly and maintaining a healthy body weight, the risks of developing diseases such as these are greatly reduced (fig 2.140).

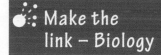

Make the link – Biology

Arteries have thick muscular walls and carry blood away from the heart towards other organs (see page 190).

Fig 2.139 *Avoiding salty foods can help to keep blood pressure in a healthy range*

GO! Activities

Activity 2.7.1 Working in pairs

Working in pairs, pick one of the following diseases and make an A4 size poster on coloured paper to display information about causes, symptoms, prevention and treatment.

- Atherosclerosis
- Heart attack
- Stroke
- Type 2 diabetes

Team up with another pair and look at their poster. Using sticky notes, write down two things the other group did well and one thing they could improve on. Stick the notes onto their poster and return it to the other group.

Activity 2.7.2 Working in pairs

The following statements give details of some strategies to help the public improve their diet and increase their exercise level. Read through the statements and discuss them with a partner. Rank the strategies from best to worst and make sure you can justify your choices.

- No fatty foods will be allowed in schools.
- Food manufacturers will add vegetable purees to foods such as lasagne to reduce the calories they contain.
- Offices where people are sedentary for long periods of time will run exercise classes at lunchtime.
- All school pupils will have one hour of compulsory PE every day.

Fig 2.140 *Regular exercise improves circulation and helps to maintain a healthy body weight*

Make the link – Health and Wellbeing

In health and food technology you study the relationship between maintaining a healthy lifestyle and food. The foods we eat can have a significant impact on our health.

(continued)

- New parents will be given cooking classes to pass on healthy eating habits to their children.
- All foods will display their salt content on the front of the packaging so that consumers know exactly how much salt they are consuming.

I can:

- Describe the consequences of a high-fat diet including becoming overweight and developing atherosclerosis.

- State that atherosclerosis is caused by a build-up of fatty deposits in the arteries.

- State that a high-salt diet can lead to high blood pressure (hypertension).

- State that lack of exercise can lead to obesity and circulatory problems.

Smoking

When a person inhales cigarette smoke they allow a substance called **tar** to enter their lungs.
Tar builds up inside the lungs and can cause lung cancer.

A smoker also inhales **carbon monoxide**. This reduces the ability of the red blood cells to transport oxygen around the body.

More than 50 of the chemicals found in cigarette smoke are known to be **carcinogenic**, this means they cause cancer.

Tar also damages the **cilia**, reducing their ability to protect the lungs from damage and infection.

Smokers also inhale toxic heavy metals such as cadmium and chromium. These heavy metals can cause cancer and damage the organs of the body.

Cigarettes contain many different poisonous substances, for example arsenic and hydrogen cyanide.

Smoking increases the risk of developing narrowed arteries. This condition is known as **atherosclerosis** and reduces blood flow to the organs of the body.

Cigarette smoke contains a chemical called benzene. Benzene is a carcinogen, associated with leukaemia.

Smoking increases the risk of developing heart disease.

Nicotine is a chemical found in tobacco that is inhaled in cigarette smoke. This is the chemical that causes **addiction**.

Alcohol

Regularly drinking alcohol can increase the risk of developing a **stroke**. This is because 'binge' drinking can **raise blood pressure**.

Heavy drinking can also trigger **type 2 diabetes**, meaning the body does not respond to insulin. This causes high blood sugar levels, which can be hazardous to health.

Drinking a large quantity of alcohol over a short period of time can cause **alcohol poisoning**. Symptoms include confusion, vomiting and seizures. In severe cases it can cause coma and death.

Regularly drinking alcohol can increase the risk of **atherosclerosis** and suffering a **heart attack**.

Drinking large quantities of alcohol regularly can damage the liver and lead to a **cirrhosis**. Cirrhosis is irreversible scarring of the liver, which impairs liver function. Eventually the liver stops working, and a transplant is required.

Heavy drinking can also affect a person's **mental health**. For example, drinking large quantities of alcohol regularly can increase the risk of depression.

Drinking alcohol is known to increase the risk of mouth, oesophageal, breast and liver cancer.

Short-term effects of drinking alcohol include blurred vision, balance problems and impaired judgement.

GO! Activities

Activity 2.7.3 Working in groups

Imagine you are part of a healthy lifestyle campaign. It is your job to inform the public about the dangers of smoking or drinking alcohol. Decide on the format of your campaign. You can choose from:

- a television advert
- information leaflet
- poster
- slideshow presentation

Work in a group and present your campaign to the rest of the class when you are finished.

Activity 2.7.4 Working individually

Write a discursive essay of about 200 words that looks at whether diseases caused by smoking and alcohol should be treated for free by the NHS. Your essay should be laid out in the following format:

Introduction – introduce the issue

Arguments for – give two or three arguments for treating diseases caused by smoking and alcohol being treated for free

Arguments against – give two or three arguments against treating diseases caused by smoking and alcohol being treated for free

Conclusion – summarise your essay and give your own opinion.

I can:

- Describe some of the consequences of smoking cigarettes including cancer, reduced ability of the blood to carry oxygen and atherosclerosis.

- Describe some of the consequences of drinking large quantities of alcohol regularly, including alcohol poisoning, cirrhosis of the liver and increased risk of stroke, type 2 diabetes and atherosclerosis.

Unit 2– Assessment

Section A

1. The diagram below shows a type of tissue found in a plant.

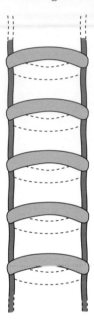

Which line in the table correctly identifies this tissue and its function?

	Tissue name	Function
A	Xylem	Transports water
B	Phloem	Transports sugar
C	Xylem	Transports sugar
D	Phloem	Transports water

2. The diagram below shows some cells as they appear when viewed through a microscope.

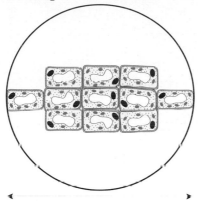

Field of view: 600 micrometres

What is the average length of the cells?

A 60

B 120

C 200

D 300

3. Which part of the brain controls heart and breathing rate?

A Spinal cord

B Cerebellum

C Medulla

D Cerebrum

4. The diagram below shows the structure of a flower.

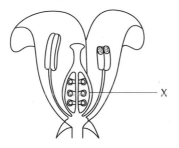

Which line in the table below correctly identifies X and the gamete it produces?

	Name of X	Type of gamete produced
A	Anther	Male
B	Anther	Female
C	Ovary	Male
D	Ovary	Female

5. The bar chart shows the results of a survey into the heights of heather plants on an area of moorland.

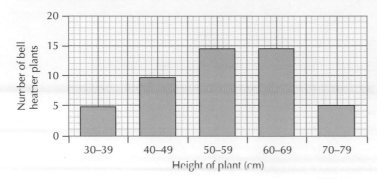

The percentage of plants with a height greater than 69 cm is

A 5%

B 10%

C 15%

D 20%

6. In humans, the allele for free earlobes is dominant to the allele for attached earlobes. Two parents both have free earlobes. Their child has attached earlobes.

What is the best explanation for this pattern of inheritance?

A The parents are heterozygous for the earlobe alleles.

B The parents are homozygous for the earlobe alleles.

C One parent is homozygous, and the other is heterozygous, for the earlobe alleles.

D The child has inherited the attached earlobes directly from a grandparent.

Questions 7 and 8 refer to the diagram below showing a cross section through a leaf.

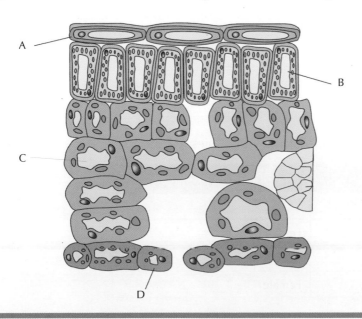

7. Which letter is pointing at a spongy mesophyll cell?

8. What is the function of cell B?

   A It allows light to pass through to the next layer.

   B It is the main site of photosynthesis.

   C It is waterproof to minimise water loss.

   D It controls whether the pores of the leaf are open or closed.

9. Which structure in the diagram below identifies the blood capillary?

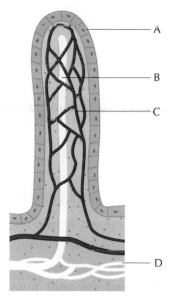

10. A group of students investigated the effect of increasing carbon dioxide concentration in the air on breathing. The volume of air inhaled by one student was measured as the carbon dioxide concentration in the air increased. The results are shown in the graph below.

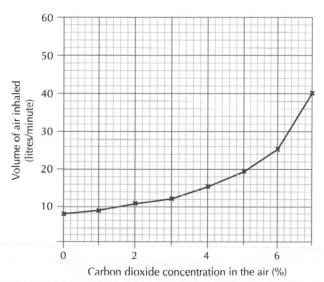

When the carbon dioxide concentration in the air is increased from 1% to 6%, the volume of air inhaled increases by

A   9 litre/minute

B   16 litre/minute

C   20 litre/minute

D   25 litre/minute

Section B

1. The table below shows diagrams of two different cell types. For each cell, explain how its structure is related to its function.

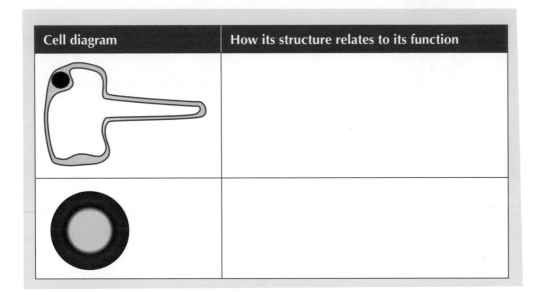

Cell diagram	How its structure relates to its function

2

2. **a)** Name the type of glands that produce hormones

_____ 1

b) Describe how hormones travel from the glands where they are produced to the place where they have an effect.

_____ 1

c) Discuss the cause and treatment of either type 1 or type 2 diabetes.

_____

_____

_____

_____ 3

3. **a)** Give the function of a reflex action.

    _____ **1**

    b) Complete the table below by placing an R next to the statement if it is a reflex action.

Response	
Swallowing when food touches the back of the throat	
Seeking shade in hot weather	
Jumping at a loud noise	
Running cold water over a burnt hand	

    ?

    c) The diagram below shows the flow of information from a receptor to an effector in a reflex action.

    Complete the diagram by inserting the names of the missing neurons in the correct order.

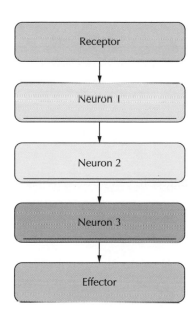

    2

4. The following diagram shows the production of sex cells and the process of fertilisation in humans. The numbers indicate the number of chromosomes in the cells.

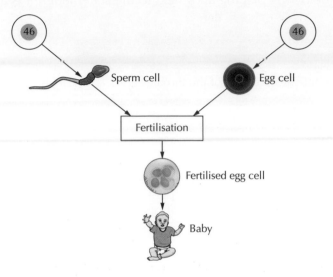

a) Give the location of sperm production in humans.

_____ **1**

b) State the term used to describe the chromosome complement of sex cells.

_____ **1**

c) Complete the diagram above to show the number of chromosomes present in each sex cell and the fertilised egg cell. **1**

d) State the term used to describe a fertilised egg cell.

_____ **1**

5. In pea plants, tall height (T) is dominant to dwarf height (t). Two pea plants each with the genotype Tt were crossed.

    a) Copy and complete the Punnett square below.

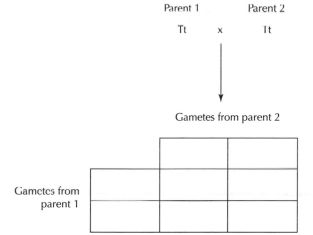

Parent 1   Parent 2

Tt   x   Tt

Gametes from parent 2

Gametes from parent 1

2

    b) State the expected ratio of tall to dwarf pea plants in the offspring.

        _____tall : _____ dwarf   **1**

    c) The cross actually produced 52 tall plants and 13 dwarf plants. Express this result as a simple whole-number ratio.

        _____tall : _____ dwarf   **1**

6. The diagram below shows the human heart.

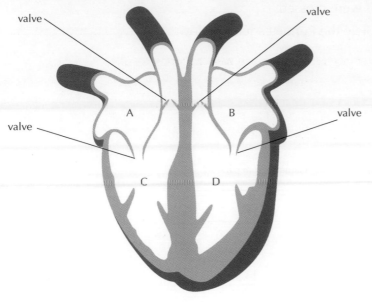

a) Name chamber C.

_____ **1**

b) State the function of the heart valves.

_____ **1**

c) Name the blood vessels that supply the heart with oxygen.

_____ **1**

7. A student carried out an investigation into the relationship between running speed and heart rate. She used a treadmill to run at a specific speed for 1 minute and measured her heart rate immediately after. She rested for 2 minutes before running at the next speed.

The results of the investigation are shown below.

Running speed (km/hour)	Heart rate (beats per minute)
0	70
2	77
4	82
6	95
8	112
10	143
12	175

a) Why was it good experimental practice to rest for 2 minutes in between each test?

_____ **1**

b) Complete the line graph below using the student's results.

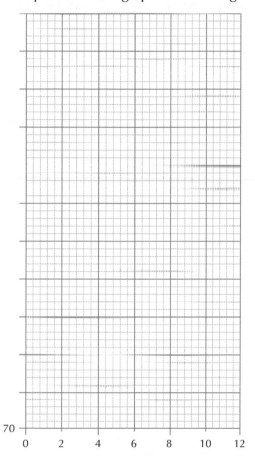

70

0   2   4   6   8   10   12

3

c) Calculate the percentage increase in heart rate as the student's speed increased from 0 km/h to 12 km/h.

_____ % **1**

d) Describe the relationship between running speed and heart rate.

_____

_____ **1**

e) How could the student make her results more reliable?

_____ **1**

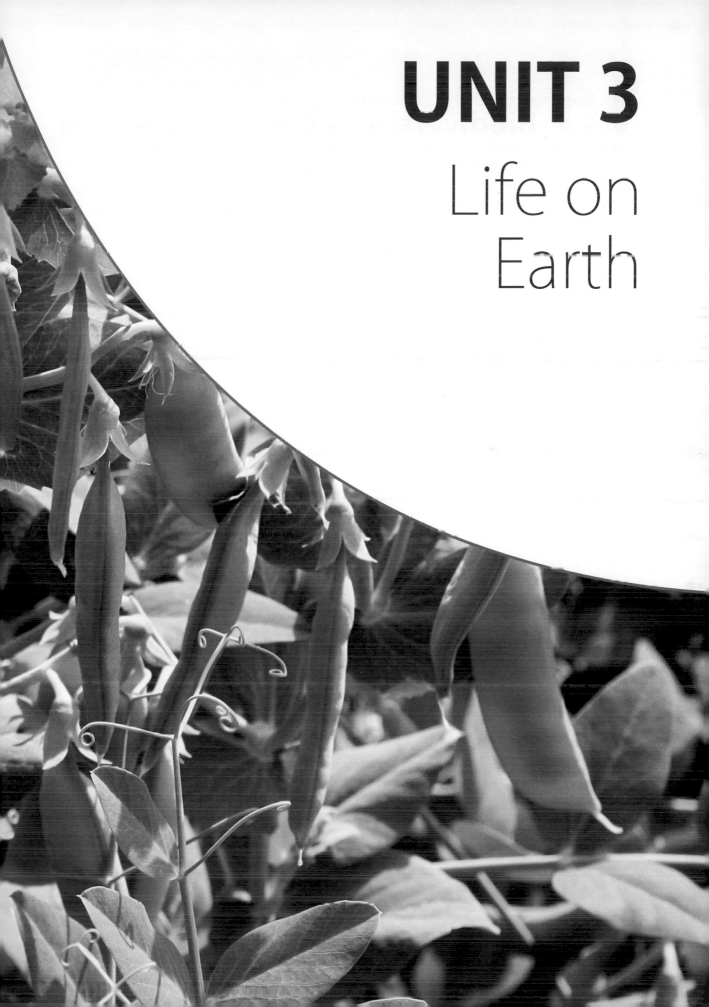

UNIT 3
Life on Earth

16 Biodiversity and the distribution of life

Biomes

The study of living organisms in their environment is called ecology. Ecologists are scientists who study the interactions between different types of organisms and between organisms and their environment.

There are **various regions** on our planet. Each of these environments is called **a biome** (fig 3.1). A biome is a geographical region of the planet that contains distinctive communities of **flora** and **fauna**, and each biome is characterised by a distinctive **climate**.

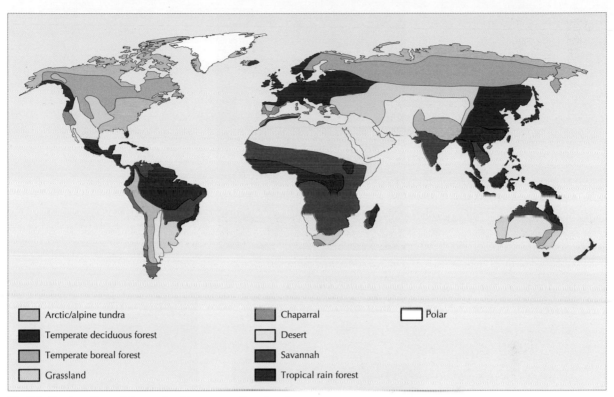

Legend:
- Arctic/alpine tundra
- Temperate deciduous forest
- Temperate boreal forest
- Grassland
- Chaparral
- Desert
- Savannah
- Tropical rain forest
- Polar

Fig 3.1 *The location of the major biomes on the planet*

Flora is the name given to the characteristic types of plants found in the biome. Fauna are the characteristic types of animals found in the biome. Ecologists consider biomes to be a group of connected ecosystems.

Biomes are very important to study so that they can be conserved and protected. Human activities have drastically altered biomes.

Global distribution of biomes

Various regions of our planet can be classed as belonging to similar biomes. Biomes are **distinguished** by their climate, flora and fauna.

Biomes can be on water or land. Five major types of biomes are:

- Forests (fig 3.2)
- Deserts (fig 3.3)
- Grassland (fig 3.4)
- Tundra (fig 3.5)
- Aquatic (fig 3.6)

Fig 3.2 *A tropical rainforest biome shows wide biodiversity*

Fig 3.3 *Desert biomes are very dry because they are either very hot or very windy. The heat and wind cause water to evaporate quickly*

Fig 3.4 *Grassland biomes are characterised by a lack of large vegetation*

Fig 3.5 *Tundra biomes are very cold and covered in frost for much of the year*

Fig 3.6 *Animals and plants living in water biomes are adapted to living life under water*

The organisms living in each type of biome are adapted to survive in the particular conditions present. They interact with each other and with the environment.

Factors influencing the global distribution of biomes

Differences between these different biomes can be **influenced by temperature and rainfall.** It is these abiotic factors that influence the **global distribution of biomes.**

Temperature

The angle at which light from the sun hits the Earth's surface influences temperature. Not all places on Earth receive the same quantity of light. The quantity of light causes differences in temperature.

Biomes found furthest from the equator receive less light. They are the coldest biomes. The Tundra biome is an example.

Biomes found closest to the Equator receive most light. They are the hottest biomes. Tropical rainforest is an example.

The temperature on the Earth's surface becomes steadily colder when moving from the Equator towards the poles.

Rainfall

Rainfall influences the distribution of biomes. The Equator is a warm region. The high quantity of light and heat from the Sun makes the air warm. The warm air causes the evaporation of water from seas and oceans. As the air rises it cools. This causes rainfall. Tropical rainstorms occur, and biomes around the Equator receive high rainfall.

At the poles the air is much colder. Less water is evaporated from seas and oceans. Less rainfall therefore occurs.

The diagram below summarises the temperature and rainfall in various biomes (fig 3.7).

<table>
<tr><td>

Make the link – Biology

Tropical biomes have a higher number of producers than other biomes. They can therefore support a larger biodiversity than other biomes.

</td></tr>
</table>

<table>
<tr><td>

Make the link – Social studies

In hotter environments such as rainforests, nutrients are recycled by decomposition much faster than in colder environments like the tundra. How would this affect the biodiversity of plants that grow in these environments?

</td></tr>
</table>

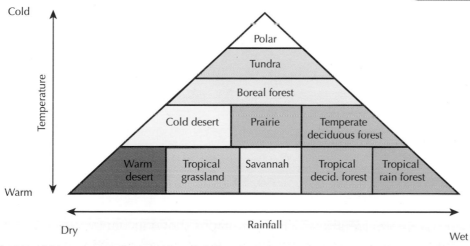

Fig 3.7 *All biomes are influenced by rainfall and temperature*

Biology in context

The higher temperatures and rainfall around the Equator result in tropical biomes supporting a wide biodiversity. This is a major concern. Some ecologists suggest that at the current rate of clearing, rainforests may be wiped out within the next 50 years. Should this happen many species of plants and animals would become extinct.

Hint

Make sure you can identify different biomes and give examples of the flora and fauna found in them.

GO! Activities

Activity 3.1.1 Working individually

(a) State the term used to describe the different types of environment found on Earth.

(b) Name three different types of biome.

(c) Explain what is meant by the terms 'flora' and 'fauna'.

(d) Explain how temperature influences the global distribution of biomes.

(e) Explain how rainfall influences the global distribution of biomes.

Activity 3.1.2 Working individually

The following graphs show the temperature and precipitation found in tundra in Russia and in rainforest in Belize.

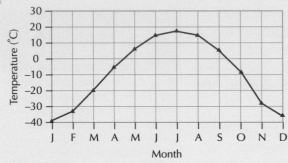

Graph 1 – *Annual temperature*

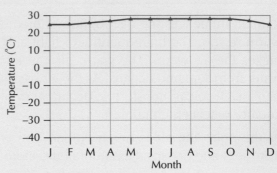

Graph 2 – *Annual temperature*

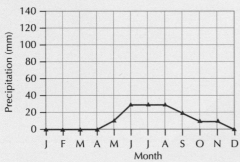

Graph 3 – *Annual precipitation*

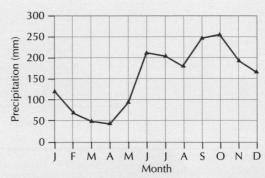

Graph 4 – *Annual precipitation*

(a) Use the temperature graphs to identify which of the graphs shows the tundra and which shows the rainforest.

(b) Describe the temperature found in each of these biomes throughout the year.

(c) In which month is there the greatest difference in temperature in the biomes?

(d) Give the month that a plant living in the tundra may also survive in the rainforest. Explain your reasoning.

(e) Use the precipitation graphs to identify which of the graphs is from the tundra and which shows the rainforest.

(f) State the month where there is the biggest difference in annual precipitation.

Activity 3.1.3 Working individually

Forest Biomes

(a) Copy the table below.

	Type of forest		
	Tropical	Temperate	Boreal
Location			
Temperature			
Precipitation			
Soil			
Effect of canopy on light reaching forest floor			
Flora (give number of tree species found per km^2 if possible and another plant example) Fauna			

(b) Search the internet for UCMP Berkeley. From their homepage search 'biomes' and click on 'The World's Biomes'.

(c) Complete the table.

(d) Choose another type of biome and produce a table similar to that above to summarise your findings.

Activity 3.1.4 Working in pairs

Grassland Biome

Grassland biomes can be divided into tropical and temperate grasslands. Tropical grasslands are called savannahs, and temperate grasslands are prairies and steppes.

(Continued)

Fauna found in the savannah can include giraffes, zebras and kangaroos. Fauna found in prairies and steppes can include badgers, hawks and owls.

Fauna living in savannahs can be put at risk due to poaching, overgrazing and clearing of land for crops. Those living in prairies and steppes can be put at risk due to over-grazing, ploughing of land and excess salts remaining in the soil from irrigation systems.

(a) Produce a spider diagram to summarise the information given above.

(b) Choose one of the fauna mentioned above and produce an information leaflet detailing the risk to the animal from human activities.

Your leaflet should include the following information:

- Natural habitat of the animal (include information on normal temperatures, rainfall, flora also found).
- Where the animal is found – include a map to show the distribution.
- What the risk from human activity is and how it affects the animal.
- Whether the animal is considered to be endangered.

Your leaflet should be an A4 piece of paper folded to A5 size.

Activity 3.1.5 Working individually

Aquatic Biome

The aquatic biome consists of marine and freshwater environments. Freshwater environments contain less than 1% salt. Marine environments contain much higher salt levels and can be as high as 40%. Freshwater environments can be divided into (i) ponds and lakes, (ii) streams and rivers, and (iii) wetlands.

Marine environments can be divided into (i) oceans, (ii) coral reefs and (iii) estuaries.

(a) Produce a spider diagram to summarise the information given above.

(b) Use the Internet to find examples of flora and fauna that you could include to complete your spider diagram.

Activity 3.1.6 Working in groups

In your jotter you should write a report on one of the following biomes – freshwater, marine, desert, forests, grassland or tundra.

You will need to

(a) Name the biome you have investigated.

(b) Give the location of the biome on earth.

(c) Give the types of flora found in the biome.

(d) Give the types of fauna found in the biome.

(e) State the annual temperature of the biome.

(f) State the annual precipitation of the biome.

(g) Give any other interesting facts about the biome.

Try searching for biomes on the internet. A useful website is saburchill.com. A search for 'biome' on this site should give you plenty of information.

Once you have completed your report you should share the information with the others in your group.

Each report should be 50–200 words long.

I can:

- State that a biome is a geographical region of the planet that contains distinctive communities of flora and fauna.

- State that each biome is characterised by a distinctive climate.

- State that flora is the name given to the characteristic types of plants found in the biome.

- State that fauna are the characteristic types of animals found in the biome.

- State that a biome is a group of connected ecosystems

- State that biomes can be found on land and in water.

- State that examples of biomes are forest, desert, grassland, tundra, freshwater and marine.

- State that the global distribution of biomes is influenced by abiotic factors.

- State that the abiotic factors that influence the distribution of biomes are temperature and rainfall.

Ecosystems

Ecology is the study of organisms in their surroundings. Ecosystem is a shortened form of ecological system. **An ecosystem consists of all the organisms living in a particular habitat**.

All the organisms in the ecosystem are called the community. Communities consist of a variety of plants and animals. **A population is the total number of organisms of the same species living in a habitat**. A habitat is the place where the organism lives.

Within the ecosystem there are many habitats. Each habitat possesses specific physical characteristics. These include light availability, moisture levels and temperature. For example, in a loch the temperature will be hotter at the surface of the water than at the bottom of the loch. At the top of a hill the soil will usually contain less moisture than at the bottom. The trunk of a tree facing north will receive less light than the part that faces south.

The living **organisms interact with the non-living components** of the ecosystem. Non-living components include temperature, light intensity, soil type, soil nutrients and rainfall.

A variety of ecosystems exist. The largest ecosystems are called biomes. Within each ecosystem there are a variety of habitats. The habitat is where the organism lives.

The diagram below shows a garden pond ecosystem (fig 3.8). In the pond there are different habitats – the pondwater, the soil and the plants themselves.

All the living organisms make up the community. The community of animals and plants are affected by abiotic factors. Abiotic

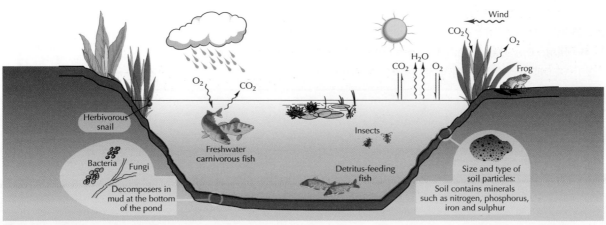

Fig 3.8 *A garden pond ecosystem*

factors are physical factors such as temperature and light intensity. The abiotic factors in the pond include oxygen availability, carbon dioxide availability, size of the soil particles, light intensity and nutrient levels.

Biotic factors are factors relating to the living organisms and include disease and predation.

Need for plants in ecosystems

Ecosystems require plants to function. Plants produce food by photosynthesis. The food is eaten by animals. Plants also release oxygen as a by-product of photosynthesis. Oxygen is used by both animals and plants for respiration. Plants provide habitats for other plants as well as animals. Plants can provide shelter from harsh weather conditions for animals.

Need for animals in ecosystems

Ecosystems require animals in order to function. Prey animals provide food for predators. Animals act as pollinators ensuring that plants can reproduce. Seeds produced by plants are dispersed by animals. Animals can eat fruits and disperse the seeds in their faeces. The fruits can attach to fur and be dispersed. Biodiversity could be reduced without animals. Dominant plants could outcompete other plants. Herbivores feeding on dominant plants allow less dominant plants to grow.

Niches

The habitat is where the organism lives. The **niche is the role the organism plays within a community**. To describe a niche it is necessary to understand that organisms interact with both the physical and biological parts of the ecosystem.

A niche **includes the use the organism makes of the resources in the ecosystem**. This could be thought of as what the organism does with the resources. It also includes the organism's **interactions with other organisms in the community**. This could be thought of as how the organism lives.

Interactions in a niche include:

- **Competition** This could be for food. In a niche, no two species of animals could feed on the same food source for a prolonged period of time. Eventually one species would be forced to move out of the habitat or face extinction.

- **Parasitism** A parasite forms a close feeding relationship with another animal. The parasite feeds from the host and

Hint

Pollinators are animals that carry pollen from one flower to another. Pollinators can be insects, birds and mammals.

Biology in context

In Scotland, only around 100 Scottish wildcats are thought to remain in the wild. The major threat to these cats is feral cats. Feral cats are domesticated cats that have returned to the wild. Feral cats spread disease to the Scottish wildcats. Feral cats also occupy the same niche as the Scottish wildcats.

239

causes it harm. It is rare for the parasite to kill the host. If it did it would lose its food source.

- **Predation** Predators eat other animals to stay alive. Predation is different from parasitism as the food source is killed. The number of predators is found to increase if the number of prey organisms increases.

- **Light** Plants compete for light to photosynthesise. Some plants grow tall to obtain maximum light. Others spread over a wider area but grow closer to the ground. Spreading plants reduce competition for light as they out-compete other plants for light.

- **Temperature** Bears are animals that have a wide distribution (fig 3.9). A polar bear would not be able to live at the Equator, as it would be too hot. Even though it may be able to exploit the niche, the climate prevents it from doing so.

- **Nutrient availability.** Decomposition occurs more quickly in hotter temperatures than in colder temperatures. Nutrients are returned to the soil more slowly in colder temperatures.

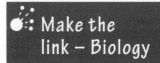

Make the link – Biology

You will learn more about adaptations that organisms evolve to allow them to live in their environment on page 307.

Make the link – Health and Wellbeing

You may have learned the importance of having a balanced diet that includes all the nutrients needed for growth. Why is it important that ecosystems contain nutrients?

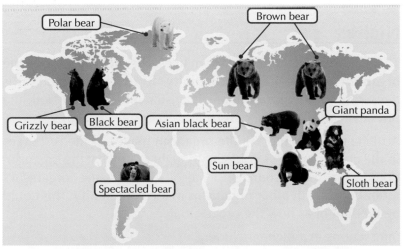

Fig 3.9 *Bears are widely distributed. They are unable to exploit the same niches due to differences in temperature*

🔵 Activities

Activity 3.1.7 Working individually

(a) State what is meant by the term 'ecosystem'.

(b) Explain why ecosystems require animals and plants.

(c) State what is meant by the term 'niche'.

(d) Give three examples of interactions organisms make with other organisms in their niche.

(e) Give three examples of resources used by organisms in their niche.

Activity 3.1.8 Working individually

The following diagram shows the distribution of two species of barnacles on a rocky seashore in Scotland (fig 3.10).

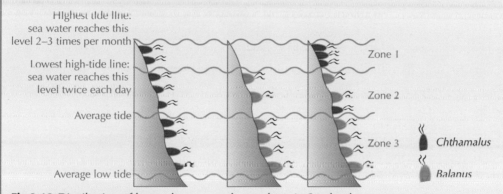

Fig 3.10 *Distribution of barnacles on a rocky seashore in Scotland*

(a) State the zones (habitats) that *Chthamalus* and *Balanus* can exploit when they are not in competition with each other.

(b) State the zones (habitats) each of these barnacles can exploit when they are in competition.

(c) Give the name of the barnacle best able to exploit the niche. Explain how you reached your conclusion.

(*Continued*)

Activity 3.1.9 Working individually

The following graph shows the distribution of red and grey squirrels in Scotland over a ten-year period (fig 3.11).

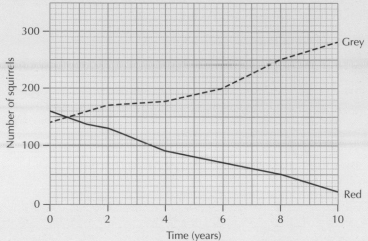

Fig 3.11 *Distribution of red and grey squirrels in Scotland*

(a) State the only time that the numbers of red and grey squirrels were the same.

(b) State the ratio of red squirrels to grey squirrels in year 1 **and** in year 10.

(c) Calculate the percentage increase in grey squirrels over the ten-year period.

(d) Calculate the percentage decrease in red squirrels over the ten-year period.

(e) State the expected number of red squirrels if the same survey was carried out in ten years' time. Explain.

Activity 3.1.10 Working in groups

Caledonian forests

Scotland was once covered in woodland known as the Caledonian forest. The forest was formed at the end of the last ice age. Less than 1% of the original native forest survives today. This is considered a problem as the forests are the habitat for many rare plants and animals.

(a) On the internet, find the 'Trees for life' website.

(b) Investigate one of the following organisms using the website:

   1. Scottish wildcat
   2. Red squirrel
   3. Red grouse
   4. Scottish crossbill
   5. Bracken

(c) Produce a slideshow presentation to share with your class. Include in the presentation:

   1. The worldwide distribution of the organism you have chosen.
   2. The distribution of the organism in Scotland.

Activity 3.5.9 Working individually

The following table shows the results of an investigation into DDT and bioaccumulation in a food chain.

Organism	DDT concentration (ppm)
Phytoplankton	0.000003
Zooplankton	0.04
Minnow	0.5
Needle fish	2
Osprey	25

(a) Draw a food chain for the organisms present.

(b) State the percentage increase in the DDT concentration between the minnow and the osprey.

(c) State the ratio of DDT in zooplankton compared to the osprey.

(d) Give the number of times more DDT is present in the tissue of needle fish than in zooplankton.

I can:

- State that a pesticide is a chemical used to kill organisms that are in competition with the crop plant.

- State that there are several types of pesticides that are specific in what they kill.

- State that herbicides kill plants, fungicides kill fungi, insecticides kill insects and bactericides kill bacteria that are in competition with the crop plant.

- State that some pesticides sprayed onto crops accumulate in the bodies of organisms over time.

- State that bioaccumulation means that the level of pesticide found in the body of organisms increases as the position of the organism in the food chain increases.

- State that DDT was sprayed onto plants and entered the food chain when the contaminated plants were eaten.

- State that DDT is a powerful insecticide that is now banned due to its harmful effect on biodiversity.

- State that DDT caused the thinning of eggshells, meaning that chicks hatched before they were strong enough to survive.

Pollution

Pollution is the addition of substances to the environment that cause harm to organisms. Human activities produce pollutants in a variety of ways. The pollutant is the substance that causes harm. Pollution can cause harm to all parts of ecosystems including land, air and water. Examples of human activities that pollute our environment can be seen in figure 3.91.

Raw sewage from an outlet pipe is being washed back onto the beach

Substances may leak into the soil. When the rubbish is covered, methane may be produced

Chemicals from the factory may pollute drinking water

Smoke from industry contributes to acid rain

Fig 3.91 *Humans pollute their environment in many ways*

Indicator species

Indicator species are organisms that **indicate levels of pollution or environmental quality**. They do this by their **presence, absence** or abundance in the environment. Examples of indicator species are:

- **Freshwater invertebrates:**
  These can indicate levels of pollution in fresh water. This is because they have different levels of sensitivity to dissolved oxygen concentrations. Fresh water can become polluted with organic waste such as sewage. Organic waste acts as a food supply for aerobic bacteria. The bacteria multiply and use up oxygen.

The species of invertebrates living in water depends on the level of pollution. Unpolluted water contains lots of oxygen. Polluted water contains lower levels of oxygen.

If the numbers and species of invertebrates living in fresh water change, this could be an early indication of pollution.

Name of indicator species	Level of water pollution	Oxygen concentration present in the water
Mayfly & stonefly nymphs	Lowest	Highest
Fresh water shrimp & caddis fly larvae	Low or medium	High or medium
Bloodworm & waterlouse	Medium or high	Medium or low
Rat-tailed maggot & sludgeworm	Highest	Lowest

Fig 3.92 *Lichens are sensitive to air pollution*

- **Lichens:**

  These can indicate levels of air pollution. This is because they have different levels of sensitivity to sulphur dioxide concentration (fig 3.92).

Air can become polluted with sulphur dioxide gas released from factories and power stations when fossil fuels are burned. Lichens are divided into four groups.

1. Crusty lichen – most tolerant to sulphur dioxide pollution
2. Leafy lichen
3. Shrubby lichen
4. 'Hairy' lichen – least tolerant to sulphur dioxide pollution.

In areas that have air pollution only crusty lichens will be found. In areas where the air is unpolluted hairy lichens as well as other types will be found.

Make the link – Biology

In Unit 3 Chapter 18 (pages **280–294**), you learned how to sample biotic and abiotic factors to show how they affect where organisms live. You now know that organisms can also be sampled to indicate levels of pollution.

Biology in context

Sampling freshwater invertebrates could be damaging to the ecosystem. Removing an organism from its habitat may cause it stress and lead to death. This may be a risk worth taking to ensure other organisms survive. The River Clyde in Glasgow was thought to be heavily polluted due to the long history of industrial use. However, in 2011 SEPA published a report that indicated there are now many species of fish flourishing in the river.

Make the link – Mathematics

You have learned how to draw graphs and work out ratios and percentages. In biology, we make use of these skills to work out relationships between organisms and their environments.

Hint

Remember pollution is anything added to the environment that causes harm. Industries and power stations could pollute water by releasing clean water that was very hot. The pollution would be thermal pollution. This occurs when the water released is not the same temperature as that taken in. This can lead to stress and the death of organisms.

Activities

Activity 3.5.10 Working individually

(a) Explain what is meant by the term 'pollution'.

(b) Give three examples of ecosystems that can be harmed by pollution.

(c) Describe what is meant by the term 'indicator species'.

(d) Predict the level of pollution and oxygen concentration in a river if mayfly larvae are found.

(e) State the pollutant to which lichens are sensitive.

Activity 3.5.11 Working individually

The following table shows the results from a water-pollution survey carried out on two different streams.

Type of invertebrate found	Number of invertebrates found	
	Stream 1	Stream 2
Mayfly larva	0	8
Caddis fly larva	0	17
Freshwater shrimp	2	78
Water louse	12	36
Bloodworm	40	8
Sludge worm	96	3

(a) Collect a piece of A5 graph paper. Draw a bar graph of the results in the table. Ensure all the data are on one graph.

(b) State the ratio of sludgeworms in stream 2 compared to those in stream 1.

(c) State the total number of organisms found in stream 1.

(d) State the percentage of sludgeworms in stream 1.

(e) State the total number of organisms found in stream 2.

(f) State the percentage of sludgeworms in stream 2.

(g) State the number of times more water lice were found in stream 2 compared to stream 1.

(h) One stream is more polluted than the other. Suggest which stream was the most polluted. Explain your choice.

(i) Explain why sludgeworms were found in both streams.

(Continued)

Activity 3.5.12 Working individually

You could carry out an investigation into the levels of pollution in two different freshwater environments. The following web link explains a possible procedure to follow: http://www.nuffieldfoundation.org/practical-biology/monitoring-water-pollution-invertebrate-indicator-species Remember to be safety-conscious if you carry out this investigation. Do not go alone.

Activity 3.5.13 Working individually

The following link contains a video that explains the way lichens can be used to indicate air pollution:

http://www.nhm.ac.uk/nature-online/life/plants-fungi/lichens-pollution/index.html

(a) Watch the video.

(b) Produce a summary note you could give a friend who has not viewed the video. The summary note should be 50–100 words in length.

(c) After watching the video you could take part in the OPAL air survey.

I can:

- State that pollution is the addition of substances to the environment that cause harm to organisms.
- State that an indicator species tells us something about the level of pollution and environmental quality.
- State that indicator species indicate levels of pollution in the environment by either their presence or absence.
- State that freshwater invertebrates and lichens are indicator species.
- State that polluted water will have low oxygen concentrations.
- State that polluted air may have high sulphur dioxide concentrations.
- State that the presence of mayfly and stonefly larvae indicates that freshwater has a low level of pollution and a high level of oxygen in the water.
- State that the presence of only rat-tailed maggots and sludgeworms indicates that freshwater has a high level of pollution and a low level of oxygen in the water.

- State that the presence of hairy lichens indicates that there is a low level of sulphur dioxide in the air.
- State that the presence of only crusty lichens indicates that there is a high level of sulphur dioxide in the air.

Biological control

The use of intensive farming methods has been shown to have a harmful effect on the environment. **Biological control and GM crops may be alternatives to the use of fertilisers and pesticides** in farming.

Biological control is seen as an alternative to using pesticides. It relies on natural predation instead of chemicals. A natural predator that feeds on the pest is released. The number of pests is reduced. An added bonus is that the cost of this control method is relatively cheap after the initial setup costs. Another benefit of using predators to combat pests is that it is a natural method and can be used in organic farming. Predators will remain in the area as long as there is food available.

Use of biological control

Many farmers use chemical pesticides to control pests. There are many disadvantages to this method.

- Chemical pesticides are non-specific. This means that beneficial insects may be killed. These insects might act as pollinators.

- Pest species may become resistant to the pesticide. All the pests in the area will eventually survive as the resistance is passed on to offspring.

- Chemical pesticides may enter the food chain. Bioaccumulation may occur and biodiversity may be reduced.

- Spraying pesticides is a health risk to those living near agricultural land.

Advantages of biological control

- The predator that is introduced is specific to the pest. This means that biodiversity may not be reduced.

3. The physical and behavioural characteristics of the organism.

4. The ecological niche (relationships) these organisms are involved in.

5. The conservation status of the organism, if that is appropriate.

6. Make your presentation 6–8 slides long.

Activity 3.1.11 Working in pairs

12% of Scotland is moorland. Moorland is land that is not farmed or covered in forest. Most of the moorland is covered in heather. Heather is important to the survival of red grouse.

(a) On the internet, go to cairngormsmoorlands.co.uk. Under the tab 'Moorland Introduction' select 'Facts & Figures'.

(b) Give examples of habitats that are included as moorland.

(c) Give examples of areas in Scotland that are considered to be moorland.

(d) Give reasons why the numbers of grouse in Scotland's moorland may be low.

Activity 3.1.12 Working in pairs

Ptarmigans are birds that are residents of Scotland. They breed in the highest mountains of the Highlands. They have foot feathers.

(a) Perform an internet search for the 'Welcome to Scotland' website. Click on the 'About Scotland' tab and then the 'Scottish Birds' tab. Then click on the Ptarmigan.

(b) State the colour of ptarmigans' feathers in both summer and winter. Explain why the change of feather colour helps these birds to survive.

(c) State how human activities have contributed to the decrease in ptarmigan numbers, especially in winter.

(d) Explain how ptarmigans' foot feathers help these birds to survive in harsh winter conditions.

Activity 3.1.13 Working in pairs

Red deer have been living in Scotland since the last ice age. They are herbivorous animals.

(a) Produce a slideshow presentation including the following points:

1. The habitat that they exploit

2. The numbers that are found in Scotland

3. The lifestyle of red deer

4. Their niche

5. Methods for controlling red deer numbers

(b) The presentation should be between 6 and 8 slides.

(c) The presentation should last 5 minutes.

I can:

- State that a population is all the organisms of one type of species in the ecosystem.
- State that a community is all the plants and animals living in the ecosystem.
- State that the habitat is the place where the organisms live.
- State that the community and habitat are influenced by non-living or physical factors called abiotic factors.
- State that an ecosystem is made up of one or several habitats and the community of organisms that live there.
- State that there are many habitats in an ecosystem.

- State examples of abiotic (non-living) factors that interact with the community include oxygen concentration, light intensity, temperature and pH.
- State that ecosystems need plants as the source of food and oxygen.
- State that ecosystems need plants as habitats.

- State that ecosystems need animals to disperse seeds.

- State that a niche is the role an organism plays in an ecosystem.
- State that the niche is the use the organism makes of the resources in the ecosystem and the interactions it makes with other organisms in the community.
- Give examples of the types of interactions made by organisms and other organisms in the community including competition, parasitism and predation.
- Give examples of the resources that organisms make use of in the ecosystem, including light, temperature and nutrient availability.

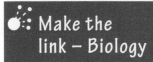

Make the link – Biology

You will learn how to sample abiotic factors in an ecosystem in more detail on page 291.

Biotic factors and biodiversity

Biodiversity in an ecosystem is affected by a variety of **factors**. These factors can be classified as **biotic** and **abiotic**.

Factors classed as biotic are of living origin. Examples of biotic factors are **predation**; **grazing**; competition (for food, space, shelter or mates); light; and disease.

Predation

A predator is an animal that hunts and kills another organism for food. Predators are carnivores. Predators kill their prey. Herbivores are sometimes considered to be predators. This is because they obtain their food by eating plants. If herbivores eat too much of a plant they may kill it.

Biodiversity can be maintained by predation. Predators keep the numbers of herbivores down. This results in less grazing. As a result there may be increased plant biodiversity.

Grazing

Animals graze on plants. Over-grazing occurs when the herbivore population is high. More animals feed on the limited plants in the habitat. This can result in plant species disappearing from the habitat. Biodiversity is reduced.

Under-grazing occurs when the herbivore population is reduced. Fewer plants are eaten. The natural assumption would be that biodiversity is increased by under-grazing. This is not the case. Biodiversity is actually reduced.

This is because dominant plants can outcompete less dominant plant species and hold them in check. This means that they do not get the chance to grow.

Moderate grazing can increase biodiversity. Grazing allows the less dominant plants to survive. This is because the reduction in dominant species gives the less dominant plants more light and soil nutrients (fig 3.12).

Biotic factors have less effect when the population size is small and the habitat contains enough resources to support the population. When the population density increases, the effect of the biotic factors increases.

Fig 3.12 *Moderate grazing increases plant biodiversity. Over- and under-grazing reduces biodiversity*

Effect of biotic factors on population

Biotic factor	Effect on population	Reason
Competition for food	Population decreases	As the population increases, the competition for food increases. Animals starve and may die. To avoid death they may move out of the habitat.
Competition for space	Population decreases	As the population increases, the competition for space increases. Animals and plants do not obtain the resources they require.
Competition for shelter	Population decreases	As the population increases, the competition for shelter increases. Animals could be exposed to harsh conditions. Some may die.

(Continued)

Biotic factor	Effect on population	Reason
Competition for mates	Population decreases	As the population increases, the competition for mates increases. Only the animals best suited to the environment mate. Fewer offspring are produced.
Competition for light	Population decreases	As the population increases, the competition for light increases. As the habitat becomes overcrowded plants struggle to obtain enough light to photosynthesise. Plants may die.
Predation	Population decreases	As the population increases, predation increases. When population size increases there are more prey for predators. More predators move into the area.
Grazing	Population decreases	When the population of plants increases there is more food for herbivores. More herbivores move into the habitat and eat the plants, decreasing the plant population.
Disease	Population decreases	As the population increases, the spread of disease increases. When the population size increases, animals and plants live closer to each other. This makes it easier for disease to spread. More organisms die.

It is important to realise that:

- When the population is smaller, the effect of the biotic factors is reduced and so the population may increase as a result.

- These factors may not affect the population on their own. Two or more factors may affect a population at the same time.

Predator–prey interactions

The effect of biotic factors on population size can be seen using a classic example of predator–prey interactions. In this case, the Snowshow hare is the prey of the Canadian lynx (fig 3.13). When the hare population increased, there was a small time delay and then the lynx population increased.

However, the increased number of lynx caused the hare population to decrease. There was more predation. Due to the reduced number of lynx, the hare population recovered and the numbers increased. This then caused the predators to increase.

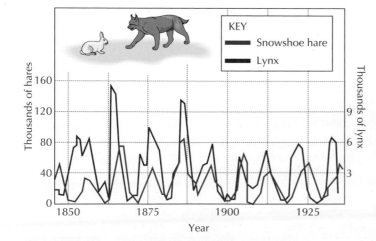

Fig 3.13 *There is a time delay between the increase in hare numbers and the increase in lynx numbers*

Abiotic factors and biodiversity

Factors that are classed as abiotic are non-living factors. They tend to be physical factors that affect the distribution of organisms in the habitat. Abiotic factors can be measured. Examples of **abiotic factors** include **pH, temperature**, moisture, light and wind speed.

pH

Sulphur dioxide and nitrogen dioxide are released when fossil fuels are burned. These gases dissolve in the water contained in clouds, forming dilute acids. When it rains, these acids fall to Earth as acid rain. Acid rain does not reduce biodiversity directly. Instead, the acid rain damages the leaves of plants. This reduces their ability to photosynthesise. Acid rain is also known to dissolve soil nutrients. This means that there are fewer nutrients available to plants.

pH can also affect the variety of fish species. Acid rain lowers the pH of freshwater. This can kill individual fish and reduce fish biodiversity. Fish that are more tolerant to a decrease in pH may survive. However, the reduced pH may lead to stress which results in a lower body weight. This results in fish being less able to compete for food and habitats.

Temperature

Carbon dioxide, nitrous oxide and water vapour are also released when fossil fuels are burned. These gases form a blanket around the Earth's atmosphere. They trap heat and cause global warming. Global warming is responsible for melting the polar ice caps. It is also thought to be responsible for causing floods.

Temperature can also affect the variety of fish species. Fish are ectotherms. Their body temperature changes with that of their surroundings. This means that they are the same temperature as their surroundings. Increased temperature results in enzyme-controlled reactions not progressing at their optimal level. Fish enzymes are adapted to function at their optimum at lower temperatures.

Increased temperature also results in increased metabolism. This means that fish require more oxygen to survive. However, increased water temperature results in less dissolved oxygen in the water. These factors all result in a decrease in fish biodiversity and are summarised in figure 3.14.

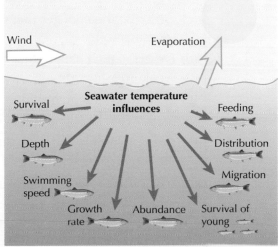

Fig 3.14 *Changes in water temperature have a dramatic effect on fish biodiversity for many reasons*

GO! Activities

Activity 3.1.14 Working individually

(a) Give two biotic factors that may affect biodiversity in an ecosystem.

(b) Explain how grazing affects biodiversity.

(c) Give two abiotic factors that may affect biodiversity in an ecosystem.

(d) Explain how pH can affect the variety of fish species in an ecosystem.

(e) Explain how temperature can affect the variety of fish species in an ecosystem.

Activity 3.1.15 Working individually

The graph below shows the relationship between a predator and prey over a period of time (fig 3.15).

(a) State the number of prey present at the beginning of the investigation.

(b) State the number of predators present at the beginning of the investigation.

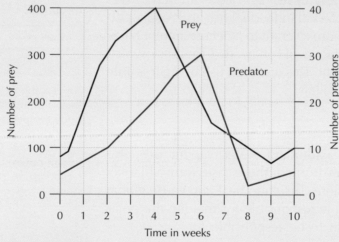

Fig 3.15 *Predator-prey interaction*

(c) Describe the changes that occur in the number of prey over the ten-week investigation.

(d) Describe the changes that occur in the number of predators over the ten-week investigation.

(e) State when the highest numbers of prey are present.

(f) State when the highest numbers of predators are present.

(g) State the number of weeks it takes for the number of predators to reach its maximum after the number of prey reaches the maximum.

(h) Explain why there is a time delay before the number of predators reaches the maximum value.

(i) Explain why the number of prey decreases from week 4.

(j) Explain why the number of predators decreases from week 6.

(k) Explain why the number of prey begins to increase from week 9.

Activity 3.1.16 Working individually

The graph below shows the effect of water temperature on percentage of fish that reach maximum size (fig 3.16).

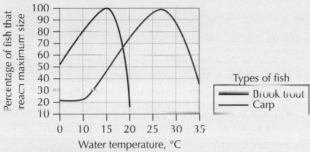

Fig 3.16 *Effect of water temperature on fish growth*

(a) State the optimum growing temperature for brook trout and carp.

(b) State the temperature range that brook trout can survive in.

(c) Describe the effect of water temperature on the growth of brook trout.

(d) State the temperature range that carp can survive in.

(e) State which of the two species would be better able to tolerate an increase in water temperature. Explain your choice.

Activity 3.1.17 Working individually

The graph below shows the biodiversity and biomass of plants in a field over a four-year period.

The data were collected to investigate the effects of grazing on biodiversity.

The field was not grazed in year 1 and 2. In years 3 and 4 sheep were grazing in the field (fig 3.17).

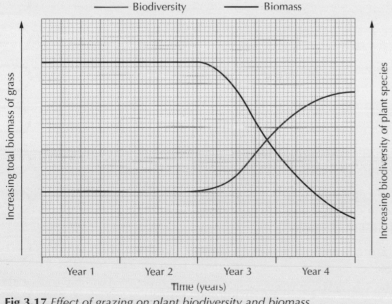

Fig 3.17 *Effect of grazing on plant biodiversity and biomass*

(a) Describe the relationship between biomass and biodiversity in years 1 and 2.

(b) Describe the effect of the grazing sheep on biomass and biodiversity.

(c) Suggest a reason why there was a time delay before the biodiversity increased once sheep began grazing the field.

(d) Explain why the biomass was greatly decreased by the end of year 4, yet the biodiversity had increased.

Activity 3.1.18 Working in groups

Barn owls are nocturnal predators. They feed on small mammals at night. This makes it difficult to know what they feed on.

Barn owls eat some food items that they cannot digest. This includes the bones of their prey. The undigested material is regurgitated. This material is known as owl pellets.

Search the internet for the Barn Owl Trust website to see an owl-pellet dissection. From this you will be able to learn more about owls as predators. On their homepage search for 'pellet' and then look at the links on the right of the page.

I can:

- State that biotic factors are those that have a living origin.
- State that abiotic factors are those that are non-living and have a physical origin.
- State that grazing and predation are examples of biotic factors that affect biodiversity.
- State that pH and temperature are examples of abiotic factors that affect biodiversity.
- State that high-intensity grazing reduces biodiversity because more plant species are eaten.
- State that moderate-intensity grazing maintains biodiversity, as it allows the less dominant plants to survive.
- State that low-intensity grazing decreases biodiversity because vigorous plants can outcompete less dominant plant species and hold them in check.
- State that high levels of predation decrease biodiversity as more animals are eaten.
- State that moderate levels of predation maintain biodiversity.

- State that low levels of predation reduce biodiversity because more dominant animals outcompete less dominant animals.
- State that acid rain damages plant leaves and indirectly reduces plant biodiversity.
- State that acid rain decreases the pH of freshwater and kills fish.
- State that acid rain may reduce biodiversity, indirectly as it can reduce the body weight of fish, making them less able to compete.
- State that an increase in temperature can reduce biodiversity, as fish enzymes do not work optimally.
- State that an increase in temperature can reduce fish biodiversity due to a lack of oxygen in the water.

Human influences and biodiversity

Human activities can also have an impact on biodiversity. Humans have reduced biodiversity in many ecosystems. This can cause the ecosystem to become unstable.

Over-exploitation

Exploitation means to use something in order to gain a benefit. Humans have exploited natural resources for hundreds of years without dramatically altering biodiversity. Problems have occurred due to increases in technology and the increase in the size of the human population. This has led to over-exploitation – taking too much from the environment. Examples include:

Over-hunting

Musk ox nearly became extinct because of over-hunting throughout the late 1900s until the 1930s. Their population has recovered due to hunting regulations and conservation programmes. Musk ox were hunted for their skin, for food and for trophies.

Over-fishing

The number of cod found in the Irish Sea and the west coast of Scotland has fallen dramatically due to overfishing. Overfishing has occurred because the catch quota has not been fixed low enough. This resulted in the population not having enough egg-producing females to replace the lost fish.

Over-grazing

In some countries, poor land-management practices have resulted in soil erosion due to over-grazing. Over-grazing results in bare soil, which is no longer suitable for growing plants.

This is a major problem, as plants add nutrients to the soil, and their roots hold the soil in place. Land that has been over-grazed becomes useless. To grow crops, more trees are felled. If this land is over-grazed, the cycle continues.

Habitat destruction

Desertification

Desertification is thought to be caused by human activities and the climatic conditions in the area. In certain areas, there are high temperatures for months on end with little rainfall. This leads to drought conditions, making it difficult for crops to grow.

If farming is the main source of income, farmers make as much use of land as possible. This leads farmers to grow as much as possible. Land is not left fallow. Fallow land is land that is rested to allow nutrient levels to recover. The land loses organic matter. This results in reduced vegetation covering the land. This leads to bare soil.

Soil erosion may occur when rain eventually falls. The rainwater washes the soil away. Soil erosion causes many problems. There is less grazing land for animals. Harvests may not give high yields. This results in people struggling to have enough food to eat. Land is therefore left fallow for shorter and shorter periods of time. All of this causes the cycle to start again (fig 3.18).

Deforestation

Tropical rainforests are habitats that have been severely damaged by human activities. In many areas, trees in the forests have been removed. The trees may be cut down to be used as building materials or burned as fuel. This contributes to air pollution. The land has then been used for farming: either as arable land or converted to pasture for animals. The land may also be converted for urban use by building homes and new roads (fig 3.19).

🔬: Make the link– Social subjects

Deforestation and desertification are problems in developing countries as people struggle to find land that is fertile enough to grow crops. How could farmers reduce the problems by better land management?

Fig 3.18 *Soil is eroded due to overgrazing. Plants are removed and no longer hold soil particles together*

Fig 3.19 *Forests are felled to make room for grazing animals or growing crops*

Tropical rainforests are able to support high biodiversity by providing many habitats and food sources for animals. Removal of the forests has led to the extinction of many species. Estimates suggest that over 100 species are lost every day due to deforestation. Removing trees results in increased competition for food between animals.

Endangered species

Organisms are said to be **endangered** when there is a risk that they may become extinct in the wild. Extinct species are no longer found on Earth. Some species are said to be extinct even though a few still remain living in protected habitats or in captivity. Threatened species are ones that could be said to be endangered in the very near future. The ten most endangered species in Britain in 2011 included Scottish wildcats and red squirrels. Species can become endangered due to:

Competition from introduced species

White-clawed crayfish are native to Britain and are endangered. Their numbers have declined due to the introduction of the North American signal crayfish in the 1970s. The American crayfish are larger and more aggressive, and carry the crayfish plague. The plague is fatal to White-clawed crayfish. Human activities such as destruction of river habitats have also led them to become endangered. American crayfish were introduced to be farmed but escaped into the wild.

Japanese knotweed is an introduced species. In the UK, over £150 million is spent every year to control this plant. The reason that the Japanese knotweed needs to be controlled is that it grows vigorously and outcompetes native species, thus reducing biodiversity. To try and control its growth, an insect called *Aphalara itadori* has been released into the wild. It feeds on knotweed.

Fig 3.20 *Japanese knotweed*

Pollution of air and water

Pollution is something that is added to the environment that causes harm. Pollution of air and water can occur.

Water pollution due to litter

Litter that ends up in water can reduce biodiversity. Animals may eat litter. This can kill the animal as their throat or stomach can be damaged.

Animals can become trapped in litter such as fishing lines or plastic bags. To free themselves, the animals may chew off the trapped part of their body. This injury makes them more prone to disease and less likely to escape from predators.

Water pollution due to organic waste

Organic waste is material that is added to water that has come from a living source. Examples include blood, sewage and oil. The organic waste provides food for bacteria. The numbers of bacteria increase and the oxygen concentration decreases as the bacteria use up more oxygen. This reduces biodiversity since larger animals such as fish are unable to survive in low oxygen concentrations.

Oil leaks from shipping receive lots of media interest. However, oil from cars also causes a lot of pollution. Oil enters the water systems after being washed of the roads by the rain. Oil prevents light reaching plants so their ability to photosynthesise is reduced. This reduces food available to animals, which may then die as a result. Animals can also die from eating the oil. The oil can be caught in feathers making it difficult for birds to fly. This makes them easier for predators to catch.

Air pollution

In Scotland there is a large biodiversity of **lichen** species. Lichens are plants. They are actually a fungus and an algae living together. The fungus provides protection for the algae. The algae provide food for the fungus. They have a symbiotic relationship. There is evidence that lichens have been on Earth for millions of years. They are important in maintaining biodiversity, as birds use them as nesting material, and they provide a home to insects. SNH states 'Scotland is important for lichens on a European and even global scale'.

Lichens are sensitive to air pollution and are referred to as an **indicator species**. An indicator species is one that gives information about the levels of pollution in an ecosystem based on its presence or absence. It is sulphur dioxide that the algal part of lichens is sensitive to.

 Activities

Activity 3.1.19 Working individually

(a) State two human activities that have resulted in habitat destruction.

(b) Name two introduced species that have caused native species to become endangered.

(c) State what is meant by the term 'pollution'.

(d) Name two pollutants of water.

(e) State the air pollutant that affects lichens.

Activity 3.1.20 Working individually

The table below shows some of the ways that humans influence biodiversity.

Human influence that causes loss of biodiversity	Organisms that have been affected
Over-hunting	
Over-fishing	
Competition from introduced species	

Copy and complete the table above.

Activity 3.1.21 Working individually

The level of nitrogen dioxide in air samples in five Scottish towns was measured.

The results are shown in the table below.

Town number	Level of nitrogen dioxide (micrograms per m³)			
	Trial 1	Trial 2	Trial 3	Average
1	61	61	61	
2	66	63	69	
3	62	66	64	
4	62	70	69	
5	68	63	64	

(a) State why three trials were carried out at each town.

(b) State the average nitrogen dioxide level in the air in each town.

(c) State the town with the highest nitrogen dioxide pollution and the town with the lowest nitrogen dioxide pollution.

(d) State the town where the results suggest that sampling technique was most reliable.

I can:

- State that exploitation means to remove resources from an ecosystem.

- State that over-hunting, over-fishing and over-grazing are ways that humans exploit the ecosystem.

- State that over-exploitation has reduced biodiversity.

- State that humans have caused habitat destruction, which has reduced biodiversity.

- State that the North American signal crayfish is an introduced species that has caused the native species of white-clawed crayfish to become endangered.

- State that Japanese knotweed is an introduced species that outcompetes native species.

- State that pollution is anything that is added to the ecosystem that causes harm.

- State that air and water can become polluted and biodiversity is reduced as a result.

- State that lichens are indicator species.

- State that lichens indicate air pollution due to sulphur dioxide.

17 Energy in ecosystems

Learning intentions

- Use the correct terminology to describe the role of organisms in food chains and food webs.
- State ways that energy can be lost as it passes along food chains and food webs.
- Describe what is represented by pyramids of numbers, biomass and energy.
- Explain why pyramids of numbers and biomass may be irregular in shape.
- State the need for nutrient cycling.
- Explain why nitrogen is needed by organisms.
- Describe the processes involved in the nitrogen cycle.
- Explain the role of microorganisms in the nitrogen cycle.
- State why competition exists in ecosystems.
- Describe interspecific competition and explain the consequences to organisms.
- Describe intraspecific competition and explain the consequences to organisms.

The role of organisms in food chains and webs

All organisms need energy to survive. The source of energy in ecosystems is the sun. Light energy is trapped by green plants and used for photosynthesis. This light energy is converted by green plants into chemical energy. They are able to make their own food so we call them producers. They are the source of food for animals.

Animals cannot make their own food. Animals rely on plants for food. Since they obtain their energy by eating other organisms, animals are called consumers.

Dead animals and plants, and their wastes, are decomposed by bacteria and fungi. The nutrients stored in their bodies are broken down and released back into the soil. These nutrients are then taken up by plants.

The following table gives definitions for the roles organisms play in feeding relationships.

Term	Type of organism	Role in ecosystem
Producers	Green plants	Photosynthesise to produce food. At death they provide food for decomposers.
Consumers	Animals	Obtain their energy by eating other organisms.
Primary consumers	Animals	Obtain their energy by eating the producers. At death they provide food for decomposers.
Secondary consumers	Animals	Obtain their energy by eating the primary consumers. At death they provide food for decomposers.
Herbivores	Animals	Only eat plants
Omnivores	Animals	Eat both plants and animals
Carnivores	Animals	Only eat animals
Top carnivore	Animals	Animals that are not eaten by any other animal. After death they provide food for decomposers.
Prey	Animals	Animals that are eaten by a predator
Predator	Animals	Animals that hunt and kill other animals
Decomposers	Microorganisms	Obtain their energy by breaking down waste, uneaten remains and dead bodies

Food chains

A **food chain** is a simple diagram showing the feeding relationship between organisms. The source of energy entering a food chain is the Sun. This is not normally shown on the diagram. All food chains begin with a green plant called the producer – the source of food. The arrows in a food chain show the direction of energy flow. They always point to the feeder in the relationship (fig 3.21).

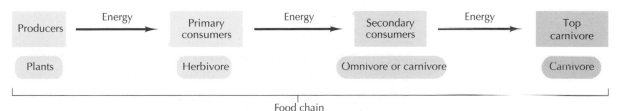

Fig 3.21 *Transfer of energy in an ecosystem*

In the simple food chain diagram on the next page, the grass is the producer. The grasshoppers are the primary consumers. They are also herbivores, as they only eat producers. They are the prey of shrews. The shrews are secondary consumers. They are the predators of grasshoppers and they are carnivores. The top carnivore is the owl. Owls are the predators of shrews, which are their prey (fig 3.22).

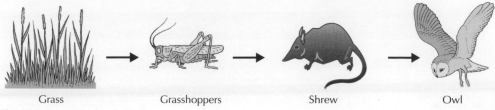

| Grass | Grasshoppers | Shrew | Owl |

Fig 3.22 *A simple food chain*

Food chains are unstable feeding relationships. There is only one source of food for the animals. If one organism is removed then the whole chain fails. They are not common in nature for this reason. The diagram below shows a freshwater food chain. If the zooplankton were removed then all the waterfleas, stickleback and pike would die, as they have no alternative food source. The phytoplankton would increase in number, as they are not being eaten (fig 3.23).

| Phytoplankton | Zooplankton | Water fleas (daphnia) | Stickleback | Pike |

Fig 3.23 *A simple, unstable freshwater food chain*

Food webs

Food webs are much more common in nature than food chains. This is because it is rare to find an organism that feeds on only one food source. Food webs are stable feeding relationships because there are alternative sources of food. A food web shows the interconnected food chains that exist in the ecosystem.

Figure 3.24 shows a simple Antarctic food web. The fish in this food web feed on three different organisms. If the carnivorous plankton were removed, the fish would survive, as they could feed on more krill or herbivorous plankton. This would also mean that the emperor penguins could also survive.

Most food web diagrams are simplified versions of the real-life situation. In the food web shown below, sea birds feed on only krill. They do also feed on other organisms, but the food web would become too complicated if all the organisms they feed on were shown.

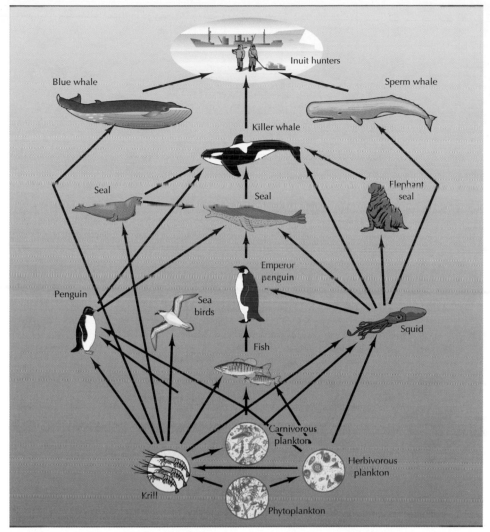

Fig 3.24 *A simple Antarctic food web*

The loss of energy from an ecosystem

In transfers from one level to the next in a food chain, 90% of energy is lost. Energy can be lost as **heat**, during **movement** and in **undigested materials**. Undigested materials include bones, hair and fur. Only **10%** of the energy available to organisms **is used** for **growth**. The ways that energy is used by organisms is shown in figure 3.25.

Energy losses from plants can include light that is not used in photosynthesis. Some of the light hitting plants can pass straight through the leaf or bounce off the surface of the leaf. Some of the light may not be of the correct wavelength to be used. Of the light energy that is used in photosynthesis, some will be lost as heat from both photosynthesis and respiration reactions.

Not all food chains and food webs rely on green plants. Green plants need light for photosynthesis. In ecosystems where there is a lack of light, such as in caves, bacteria are the source of food.

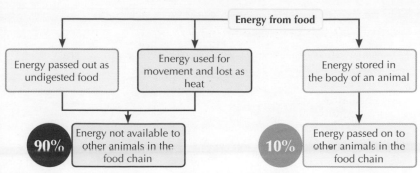

Fig 3.25 *Fate of energy flowing through a food chain*

Make the link – Biology

In Unit 1 Chapter 7, you found out that photosynthesis is vital for humans to survive. Our food comes either directly or indirectly from the process of photosynthesis. Vegans eat food only from plant sources. Vegetarians eat dairy products and so rely on animals for food as well as plants.

Make the link – Sciences

The law of conservation of energy states that energy can be neither created nor destroyed but can be converted from one form into another. What energy conversion takes place in plants?

🔍 **Hint**

Remember that the arrows in food chains and food webs always show the direction of energy flow. This means that they always point to the feeder – the organism obtaining the energy.

Some people think that only animals respire. All organisms respire to obtain energy.

Energy losses from animals include energy that is lost as heat through respiration. The majority of the energy taken in by animals is lost as heat, faeces and urine, and during movement.

GO! Activities

Activity 3.2.1 Working individually

(a) State the source of energy in all ecosystems.

(b) Explain what the arrows in food chains and webs show.

(c) Suggest why all food chains begin with producers.

(d) Give three ways in which energy can be lost from a food chain.

(e) State the quantity of energy that is lost at each stage in a food chain.

Activity 3.2.2 Working individually

(a) Copy the table below.

(b) Place examples of the organisms shown in fig 3.24 into the correct column.

Type of organism		
Producer	**Primary consumer**	**Secondary consumer**

Activity 3.2.3 Working individually

A food web found in a Scottish forest ecosystem is shown below.

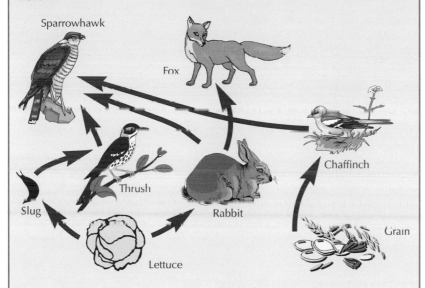

Sparrowhawk

Fox

Chaffinch

Thrush

Rabbit

Slug

Lettuce

Grain

(a) State the producer(s) in this food web.

(b) Name the top carnivore(s) in this food web.

(c) State three ways in which energy could be lost from this food web.

(d) Draw a food chain contained in the food web. The chain should consist of four organisms.

(e) Name two prey organisms found in the food web.

(f) Name two predators found in the food web.

(g) Give the organisms the sparrowhawks feed on.

(h) Explain why the rabbit numbers may decrease if the thrushes were removed from the food web.

Activity 3.2.4 Working individually

Waterfleas are the food for sticklebacks but they feed on zooplankton. Zooplankton feed on phytoplankton, which are the producers. The top carnivore is the pike, which feed on stickleback.

(a) Create a simple food chain using the information above.

(b) 100 000 kJ/m^2/year of energy is stored in phytoplankton. State how much energy will be left by the time this energy is transferred to pike.

I can:

- State that green plants are producers and that they produce the food in ecosystems.
- State that animals are consumers, as they are unable to produce their own food.
- State that primary consumers are animals that eat plants only and that they can also be called herbivores.
- State that secondary consumers that eat only other animals are called carnivores and those which eat plants and animals are called omnivores.
- State that a food chain is a simple feeding relationship that in nature is unstable.
- State that a food web is a more complex feeding relationship that consists of many interconnected food chains. This makes it more stable.
- State that if the producer were removed from the food chain, all consumers would die as they have no food.
- State that if the primary consumer were removed from the food chain, the secondary consumer would die, as it has no food. The producer may thrive, as it is not being eaten by the primary consumer.
- State that if the secondary consumer is removed from the food chain the primary consumer might thrive as it is not being eaten. However, the primary consumer might overgraze the producer and both might ultimately die.
- State that energy can be lost from food chains as heat, in movement and in undigested material.

Pyramid of numbers

A food chain shows the feeding relationship between organisms in an ecosystem. It does not tell us anything about the numbers of each of the organisms present at each feeding level. If the number of producers decreases, the consumers may die due to lack of food. A **pyramid of numbers** is a diagram that shows the **numbers of organisms present at each link of the food chain**.

The producers are always at the base of the pyramid. The top carnivore is at the tip of the pyramid. The pyramid shape is due to the decrease in numbers at each link of the food chain. The longer the bar, the higher the number of organisms. In regularly shaped pyramids, the size of the organism increases, moving along the food chain.

The following food chain (fig 3.26) can be represented as a pyramid of numbers (fig 3.27). There are most grass plants and fewest owls so a pyramid shape occurs.

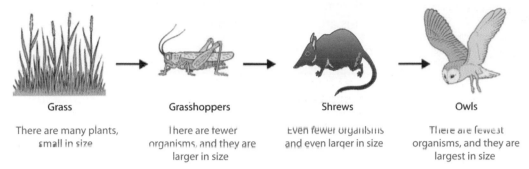

Grass	Grasshoppers	Shrews	Owls
There are many plants, small in size	There are fewer organisms, and they are larger in size	Even fewer organisms and even larger in size	There are fewest organisms, and they are largest in size

Fig 3.26 *A forest food chain*

There are some exceptions to the classic pyramid shape. These pyramids are said to be **irregular**. Some food chains begin with a single large producer, for example a tree. In this case the base of the pyramid (producer) is narrower than that of the primary consumers (fig 3.28).

Another exception is when a parasite is the top carnivore. In the food chain

lettuce → rabbit → fox

the fox may have parasites called fleas. The fleas are tiny compared to the fox. The pyramid of numbers when the fleas are included would look like this (fig 3.29).

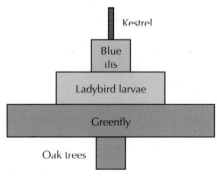

Fig 3.27 *A pyramid of numbers*

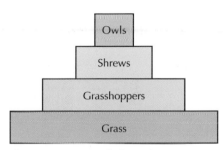

Fig 3.28 *An irregular pyramid, since the producer is a single large organism*

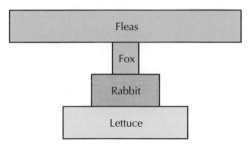

Fig 3.29 *The fleas have a much longer bar than the fox as there are many of them*

Pyramid of biomass

Because of the above exceptions, pyramids of numbers are not the best way to illustrate feeding relationships. Instead a **pyramid of biomass** is used.

A pyramid of biomass is a diagram that shows the **biomass of organisms present at each link of the food chain**. Biomass can be thought of as biological weight. The most reliable measurement of mass is dry mass. This is not used often, as organisms are killed to obtain dry mass.

The food chain

Grass → grasshopper → shrew → owl → parasites

can be redrawn as a pyramid of biomass (fig 3.30). The use of biomass instead of numbers results in the expected pyramid shape. This is because although there are many parasites they weigh much less than the owl they feed on. Since a pyramid of biomass relies upon weighing organisms a true pyramid shape is almost always obtained. There are exceptions. Pyramids of biomass can also be irregular.

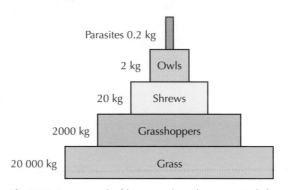

Fig 3.30 *A pyramid of biomass has the expected shape*

Zooplankton are animals that feed on phytoplankton. Zooplankton eats phytoplankton very quickly. This means the phytoplankton never have a higher biomass than the zooplankton (fig 3.31). However, there is always enough phytoplankton for the zooplankton to feed on. Phytoplankton reproduces very quickly.

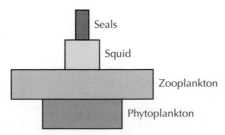

Fig 3.31 *An irregular pyramid of biomass. Phytoplankton reproduce very quickly so they are able to support the zooplankton feeding on them*

Pyramid of energy

A **pyramid of energy** is the most reliable way of representing feeding relationships in a food chain. This is because energy is always lost at each link in a food chain.

A pyramid of energy is a diagram which shows **the total energy contained within organisms present at each link of the food chain.**

Pyramids of energy are most difficult to obtain. This is because the energy content of each population has to be calculated. To do this, a sample of each organism has to be burned in a calorimeter. This gives the energy content for one member of the population. This figure has to be multiplied by the total number of organisms in the population.

A pyramid of energy is shown for the irregular pyramid of biomass. Energy is lost as heat at each link. This results in the shape always being a true pyramid (fig 3.32).

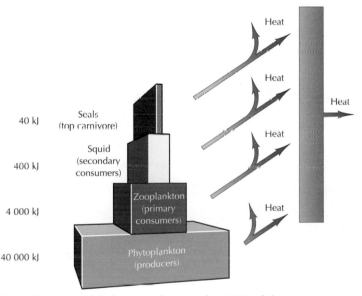

Fig 3.32 *A pyramid of energy showing that 90% of the energy at each link in the food chain is lost*

Make the link – Biology

In order to draw any of these pyramids the population has to be sampled. You will find out how to sample populations in Unit 3, Chapter 18.

Make the link – Social studies

You might have learned about human population pyramids. These pyramids show the numbers of people of each sex. They also show the age ranges of the population. What use may governments make of the information shown in human population pyramids?

🔍 Hint

Remember that heat is lost at each link in the food chain. All organisms produce heat by respiration.

GOI Activities

Activity 3.2.5 Working individually

(a) State what a pyramid of numbers shows.

(b) State what a pyramid of biomass shows.

(c) State what a pyramid of energy shows.

(d) Give a reason why a pyramid of numbers can sometimes be irregular.

(e) Give a reason why a pyramid of biomass can sometimes be irregular.

Activity 3.2.6 Working individually

In a garden, 1 rose bush provided the food for 500 greenfly. The greenfly were preyed upon by 10 ladybirds, which in turn were preyed upon by 1 blackbird.

(a) Construct a food chain from the information provided.

(b) Draw a pyramid of numbers using the information given. State whether the pyramid is a regular or irregular pyramid. Explain how you reached your conclusion (hint – consider the numbers of organisms at each link in the food chain).

(c) The rose bush was found to contain 9 000 kJ of energy at the beginning of the growing season. Ninety per cent of the energy is lost as energy is passed on to each link in the food chain.

Copy the pyramid of energy shown below and complete the energy available at each level.

Activity 3.2.7 Working individually

An investigation into brown trout in a Scottish loch was carried out. It was found that brown trout fed on dragonfly nymphs. Algae provided the food for mayfly nymphs, which were the food for dragonfly nymphs.

The investigation also measured the biomass of all the organisms mentioned above. There was found to be 1.5 g/m² of brown trout. This was 10 times less than the biomass of the dragonfly nymphs. The mayfly nymph biomass was 22 times less than the biomass of the algae, which had a biomass of 814 g/m².

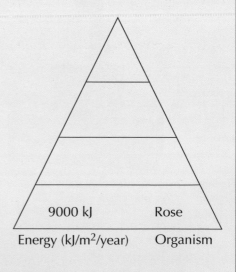

9000 kJ   Rose

Energy (kJ/m²/year)   Organism

(a) Produce a table to summarise the information given in the passage above.

(b) Explain why the biomass value decreases from one level to the next.

I can:

- State that a pyramid of numbers shows the numbers of organisms at each link in the food chain.
- State that the numbers of organisms present usually decrease, moving along the food chain.
- State that the size of the organisms usually increases, moving along the food chain.
- State that in some cases irregular pyramids of numbers are formed.
- State that a single large producer produces an irregular pyramid of numbers.
- State that parasites as the top consumers produce an irregular pyramid of numbers.
- State that pyramids of biomass and energy are less likely to produce irregular pyramids.
- State that pyramids of biomass show the mass of organisms at each link in the food chain.
- State that pyramids of biomass may produce an irregular pyramid if the producer is consumed quickly by the primary consumer.
- State that a pyramid of energy shows the energy present in the organisms at each link of the food chain.
- State that a pyramid of energy always takes a regular pyramid shape.
- State that heat is lost at each link in the food chain.

Nutrient cycling

All organisms are made from chemical elements. These elements include carbon, hydrogen, oxygen and **nitrogen**. The elements are used to produce chemicals needed by the organism. In all **ecosystems** there is a limited supply of such nutrients. If the nutrients were not recycled they would run out.

Decomposers are responsible for the recycling of nutrients. They return the nutrients stored in the dead bodies of animals and plants to the soil. They also return the nutrients stored in animal waste to the soil. The nutrients are then taken up by producers and incorporated into their tissues. Producers pass on the nutrients to consumers when they are eaten. So the cycle continues. This ensures that minerals are available in the ecosystem.

The importance of nitrogen

Plants and animals need nitrogen to make **proteins**. These proteins are **produced using nitrogen from nitrates**. Plants use nitrates to make protein. Animals eat plant proteins to make proteins in their bodies.

The nitrogen cycle

Nitrogen is a mineral that is recycled in ecosystems. The processes involved in the recycling are known collectively as the nitrogen cycle. The nitrogen cycle relies heavily on the action of **decomposers such as fungi and bacteria**. These decomposers are important as they **convert proteins and nitrogenous wastes into ammonium compounds. These are converted to nitrites and then to nitrates**.

Some leguminous plants have bacteria living in their roots (fig 3.33). The bacteria live in root nodules. The bacteria can change nitrogen gas into nitrates, which the plant then uses. Another source of nitrates in the cycle is lightning. Animals eat plants and other animals to obtain the nitrogen they need to make proteins.

Fig 3.33 *Root nodules on leguminous plants contain bacteria. The bacteria are able to convert nitrogen gas into nitrates, which can be used by the legume to produce protein*

The diagram below shows the nitrogen cycle in detail (fig 3.34). Lightning increases ammonium compounds in the soil. These can be converted to nitrates. This is represented as a dotted arrow, as it is not an everyday occurrence. An artificial way of adding nitrates to soil is to use fertilisers.

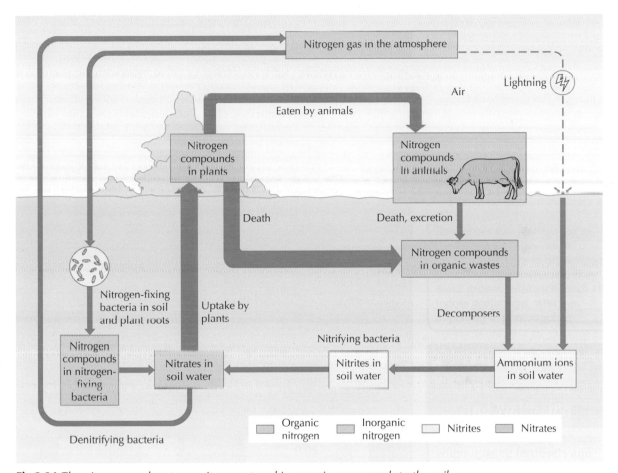

Fig 3.34 *The nitrogen cycle returns nitrogen stored in organic compounds to the soil*

The nitrogen cycle begins with larger decomposers such as earthworms and woodlice. These animals begin the process of decomposition by feeding on organic matter. Organic matter is material that has been living, such as dead animals and plants. Decomposers break down organic matter into smaller particles. This allows microscopic decomposers such as bacteria and fungi to feed on the smaller particles and break them down into their constituent nutrients.

Microorganisms and nitrogen cycling

Several microorganisms play important roles in the nitrogen cycle. Their roles are summarised in the table below.

Type of microorganism	Process carried out	Role in the cycle
Nitrifying bacteria	Nitrification	These bacteria help to increase the nitrate levels in the soil. There are two types. One type changes ammonium compounds into nitrites. The other type changes nitrites into nitrates. The bacteria use the energy released in these changes to make food. These bacteria survive in aerobic conditions.
Root nodule and free-living nitrogen fixing soil bacteria	Nitrogen fixation	These bacteria increase the level of nitrates in the soil. They are able to convert nitrogen gas into nitrates. One type of bacteria lives free in the soil. The other lives in root nodules found on leguminous plants.
Denitrifying bacteria	Denitrification	These bacteria are responsible for a decrease in the nitrate level in the soil. They change nitrates into nitrogen gas. They make the soil less fertile. These bacteria survive in anaerobic conditions.

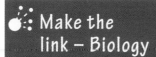

Make the link – Biology

Farmers add nitrates to the soil in the form of fertilisers. **Fertilisers supply nitrates to increase** crop **yield**. This is covered in detail in Unit 3, Chapter 20.

Biology in context

To increase nitrates in the soil in a natural way, farmers may plant legumes. These are plants such as peas, beans or clover. These plants can convert nitrogen gas into nitrates. Organic farmers do this so they can avoid using chemical fertilisers. This means that they are unable to grow a more valuable crop to sell. However, they avoid the cost of buying fertilisers.

Make the link – Ecoschools

You may be aware of the need to recycle paper. Why might we run out of resources if we don't recycle?

Hint

To help you remember the nitrogen cycle, think ANiNa – **A**mmonium **N**itr**I**tes **N**itr**A**tes

Activities

Activity 3.2.8 Working individually

(a) Name a chemical element that plants and animals need to make protein.

(b) Give the form of this chemical element that plants use to make protein.

(c) State the role of decomposers in the nitrogen cycle.

(d) Explain the role of nitrifying and denitrifying bacteria in the nitrogen cycle.

(e) Explain the role root nodule bacteria play in the nitrogen cycle.

Activity 3.2.9 Working individually

The diagram below shows part of the nitrogen cycle.

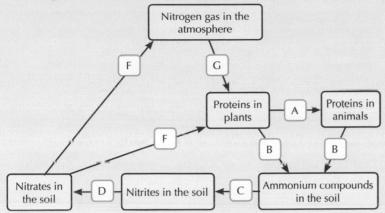

(a) Explain what is represented by arrow A.

(b) State which letter(s) represent death and decomposition.

(c) State which letter(s) represent the action of nitrifying bacteria.

(d) State which letter(s) represent the action of denitrifying bacteria.

(e) State which letter(s) represent the action of nitrogen fixing bacteria.

(f) State where nitrogen-fixing bacteria are found.

(g) Explain why nitrogen cycling is important in ecosystems.

Activity 3.2.10 Working in pairs

Nitrifying bacteria are important in the cycling of nitrogen in ecosystems. Earthworms and woodlice are important large decomposers in the nitrogen cycle.

(a) Explain how nitrifying bacteria increase the fertility of the soil.

(b) Explain why nitrogen may cycle more slowly if large decomposers were to be removed.

Activity 3.2.11 Working in pairs

An investigation was carried out using barley and pea plants.

Four pots each containing five seeds were planted. The contents of the pots are shown in the table below.

Pot number	Contents of pot		
	Seed planted	Nutrient missing from soil	Nitrogen-fixing bacteria added
1	Barley	None	No
2	Barley	Nitrogen	Yes
3	Pea	None	No
4	Pea	nitrogen	Yes

(Continued)

The seeds were allowed to grow for 4 weeks. The heights of the seedlings were recorded.

Type of plant	Barley		Pea	
Nutrient missing from soil	None	Nitrogen	None	Nitrogen
Average height of seedling (cm)	30.0	24.0	34.0	30.6

(a) State the reason for including pots 1 and 3 in the investigation.

(b) Give two variables not mentioned that need to be controlled to make the experiment valid.

(c) Suggest how the reliability of the results could be improved.

(d) Calculate the percentage decrease in the average height of the barley and pea seedlings when grown without nitrogen.

(e) State which plant grew best when nitrogen was missing from the soil.

(f) Suggest a reason for adding nitrogen-fixing bacteria to pots 2 and 4.

I can:

- State that there is a limited supply of nutrients in any ecosystem.

- State that decomposers are responsible for returning nutrients to the soil.

- State that soil nutrients are taken in by producers and incorporated into their tissues. The producers are then eaten by consumers, which incorporate the nutrients into their tissues. The nutrients are returned to the soil by the decomposition of wastes and dead bodies.

- State that plants and animals require nitrogen to produce proteins.

- State that plants such as peas and clover are called legumes. These plants contain bacteria in their roots that can convert nitrogen gas to nitrates.

- State that fungi and bacteria convert proteins and nitrogenous wastes into ammonia, nitrite and then nitrate.

- State that the nitrogen cycle is the name of the sequence of reactions that return nitrogen to the soil.

- State that nitrifying bacteria increase the level of nitrates in the soil by first changing ammonium into nitrites and then changing nitrites into nitrates.

- State that root-nodule and free-living, nitrogen fixing soil bacteria increase the level of nitrates in the soil by converting nitrogen gas into nitrates.

- State that denitrifying bacteria decrease the level of nitrates in the soil by converting nitrates into nitrogen gas.

Competition in ecosystems

Animals and plants require particular resources to stay alive. Resources needed by plants include **light**, **water**, space and soil nutrients. Resources needed by animals include **food**, **water**, shelter and mates. Organisms need to fight to obtain resources if the supply decreases.

Interspecific competition

Individuals of different species may **require similar resources in an ecosystem.** This is an example of **interspecific competition** as different species do not have exactly the same requirements. This type of competition is usually less intense than intraspecific competition when resources are in short supply. Different species do not have exactly the same need for resources.

If competition becomes very fierce, one of the species may be forced to move out of the ecosystem. This may happen when resources are extremely scarce. This happened to red squirrels and brown trout.

Red squirrels are native to Britain (fig 3.35). In 2012, there were thought to be around 120 000 in Scotland and only 140 000 in Britain. The grey squirrel was introduced to Britain in the late 19th century (fig 3.36). They are more efficient feeders than red squirrels and outcompete them. They also carry a virus that kills red squirrels. Red squirrels were forced to move out of their habitats to avoid extinction. They are now only found in areas where there are few grey squirrels.

Brown trout are native to Scotland (fig 3.37). They live in fresh water. Brown trout breed in rivers and migrate to lochs to feed. They return to rivers to breed. In 2007, brown trout were added to the UK Biodiversity Action Plan Priority Species List. One reason for this is interspecific competition from rainbow trout (fig 3.38). Rainbow trout were introduced to Scottish lochs to be farmed. Unfortunately some escaped. They are stronger and

Fig 3.35 *Red squirrels are native to Scotland. Their numbers are declining*

Fig 3.36 *Grey squirrels were introduced to Scotland. They have outcompeted red squirrels*

Fig 3.37 *Brown trout are native to Scotland. Their numbers are declining*

Fig 3.38 *Rainbow trout were introduced to be farmed. They have outcompeted brown trout*

more aggressive feeders than brown trout. They outcompeted the brown trout. Brown trout numbers reduced.

Intraspecific competition

Competition between **individuals** of the **same species** is called **intraspecific competition**. This type of competition is more intense than interspecific competition when resources are in short supply. This is because individuals of the same species **require the same resources.**

Intraspecific competition can lead to the death of the least well-adapted members of the population. They are unable to obtain all the resources they require to survive. This is an example of natural selection in action. Intraspecific competition can regulate the size of the population.

In order to reduce intraspecific competition, some species of birds have developed territorial behaviour. Robins (fig 3.39) and red grouse (fig 3.40) are examples. A territory is an area that is defended. The territory contains all the resources needed by the bird and its family. The poorer the environmental conditions, the larger the size of the territory. Having a larger territory means a larger area has to be defended. This costs more energy.

Make the link – Biology

You learned about the variety of biomes on earth (see page 230 to remind yourself). Closely related organisms are adapted to allow them to live in different biomes. This helps to reduce competition.

Fig 3.39 *Male robins defend their territory by singing a high-pitched song. They also puff out their red breast to appear larger to the intruder*

Fig 3.40 *Male red grouse defend their territory by leaping into the air. They also make a call to deter intruders*

Make the link – Social studies

You may have learned about wars that have been fought in the past. Many wars have been fought to acquire more land. Why have so many wars been fought to acquire land?

Biology in context

Grey squirrels are not only reducing the population of red squirrels, but also causing damage to woodlands. They do this by stripping the protective bark from trees.

> ### Hint
>
> Remember that 'intra' means 'within' and 'inter' means 'between'. Competition within the same species is called 'intraspecific competition', and competition between different species is called 'interspecific' competition.

GO! Activities

Activity 3.2.11 Working individually

(a) Give a definition of the term 'interspecific competition'.

(b) Give a definition of the term 'intraspecific competition'.

(c) Explain why interspecific competition is usually less fierce than intraspecific competition.

(d) Predict the effect of an accidental release of rainbow trout into a Scottish loch on the food available to brown trout.

(e) Name two birds that defend a territory. Explain why these birds expend energy to defend their territory.

Activity 3.2.12 Working individually

An investigation into intraspecific competition in mustard seedlings was carried out.

Two yoghurt pots that were wrapped in black paper were collected. 3 g of cotton wool was used in each pot instead of soil. 30 cm³ of water was added to each pot.

1 g of seeds was planted in one pot; 10 g of seeds was planted in the other. The pots were placed on a warm window ledge for 5 days.

To obtain results, the seedlings were removed from the pots and placed in an oven to dry to a constant mass.

(a) Draw a diagram to show the contents of the pots. Make sure you label your diagram.

(b) Suggest why the pots were wrapped in black paper.

(c) Suggest why cotton wool was used instead of soil as the growing medium.

(d) Give two resources the seedlings could be competing for.

(e) Explain why the pots were placed on a warm window ledge.

(f) Explain why the seeds were allowed to grow for 5 days.

(g) Name the pot where intraspecific competition was most intense.

(h) Suggest why the seedlings were dried to a constant mass before the results were obtained.

(Continued)

Activity 3.2.13 Working individually

The diagram below shows the distribution of squirrels from 1945 to 2010.

1945
■ Red Squirrel
■ Grey Squirrel
■ Both
■ None

2010
■ Red Squirrel
■ Grey Squirrel
■ Both
■ None

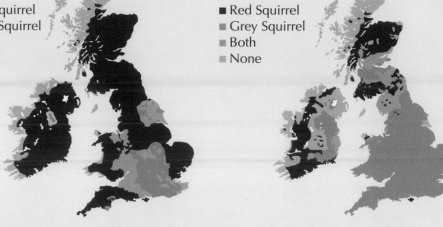

(a) Describe the change in distribution of red squirrels in Scotland between 1945 and 2010.

(b) Describe the change in distribution of grey squirrels in Scotland between 1945 and 2010.

(c) Name the type of competition that occurs between red and grey squirrels.

(d) Suggest a reason for the absence of either type of squirrels in the northwest of Scotland.

I can:

- State that individuals living in the same ecosystem may have to compete for the same resources.

- State that competition occurs when resources are in short supply.

- State that plants compete for light, water, space and soil nutrients.

- State that animals compete for food, water, shelter and mates.

- State that interspecific competition occurs between different species in the same ecosystem.

- State that intraspecific competition occurs between members of the same species in the ecosystem.

- State that intraspecific competition is more intense than interspecific competition when resources are in short supply.

- State that intense competition results in the death of the organism.

- State that intense competition may force animals to leave the ecosystem.

- State that both red squirrels and brown trout are examples of species that have been outcompeted by introduced species.

- State that some birds maintain territories to reduce intraspecific competition.

18 Sampling techniques and measurement of abiotic and biotic factors

You should already know:

- How to use sampling techniques such as quadrats and transects to sample ecosystems.
- How to use branching keys to identify organisms.

Learning intentions

- Describe methods used to make sampling techniques representative.
- Describe methods used to make sampling techniques reliable.
- Explain the need to sample habitats.
- Describe sampling techniques for plants and animals.
- Identify limitations and sources of error that might accompany the use of sampling techniques.
- Explain methods that can be used to limit the sources of error.
- Describe methods to sample abiotic factors in a habitat.
- Identify limitations and sources of error that might arise when sampling abiotic factors.
- Explain methods that can be used to limit the sources of error.
- State why keys are used to identify organisms.
- Use and construct paired statement keys to identify organisms.

Sampling

An ecologist may want to find out the type of plants and animals present in a habitat and to find out the number of different species in a habitat. It would usually be impossible to count every single animal or plant. There could be several reasons for this.

- Number of organisms: There could be too many to count without making an error.
- Time: Large numbers of organisms would take too long to count.
- Habitat damage: Certain sampling methods (e.g. pitfall traps) may result in the unacceptable death of organisms.

Sampling allows a representative picture of the types and numbers of organisms present to be determined. No sampling method will allow all the different types of animals and plants in the ecosystem to be calculated.

Sampling and Quantitative Measurements

A sample is a small representative part of the whole habitat. In ecology, samples are taken to try to find out patterns or trends occurring in habitats.

To be sure that **representative sampling** is carried out, careful planning is needed. Techniques assume that the samples taken include examples of all organisms present in the habitat. Samples need to be taken using identical sampling apparatus. This increases the validity of the sampling technique. It allows comparisons to be made.

To be representative, several samples have to be taken. The larger the area being studied, the higher the number of samples.

Samples also need to be random to be representative. This means that the experimenter should not choose where the sample is taken.

The larger the area being sampled, the more samples need to be taken.

Quantitative measurements involve counting organisms allowing comparison is to be made. Trends and patterns can be identified.

Quadrats

This technique is used to sample a variety of habitats. It is used to sample plants and animals fixed to the surface of, for example, rocks on a seashore. A **quadrat** is a square frame usually divided into smaller squares (fig 3.41). It is used to estimate a population in a habitat.

Using a quadrat

Quadrats should be dropped randomly. This can be done by the sampler turning away from the area to be sampled and dropping the quadrat over their shoulder. The plants or animals present in the quadrat are then counted.

Limitations and sources of error using quadrats

There are several limitations and sources of error.

- Organisms in the quadrat may be wrongly identified.
- Organisms in the quadrat may be wrongly counted.
- Too few samples may be taken to be representative. This is especially true when the organisms are clustered in small areas throughout the whole habitat, e.g. plants that are only found in the shade of scattered trees in an area of grassland.

Fig 3.41 *Quadrats are usually subdivided into smaller squares. Only the squares containing the organisms are counted*

Ways that the errors can be reduced

There are ways to reduce errors:

- Use a key to make sure that the organisms are correctly identified.

- If organisms are only part in the quadrat, make a rule for counting them. You could only count organisms where at least half is in the quadrat. Another method would be to count only those organisms on the top and right-hand side of the quadrat.

- If it is noticed that organisms are in clusters, increase the number of samples taken.

Using quadrats to estimate the population size

The population of an organism can be estimated by applying quadrat results to a whole habitat.

For example:

The area sampled was 25 m long and 25 m wide. The total area was 625 m².

10 quadrats were used in the area. The total number of a species of plant in the 10 quadrats was 58. This means the average number per quadrat was 5.8.

Each quadrat was 1 m² so the total population size is 625 × 5.8 = 3625 individuals.

The figure below gives more detail of the way a random quadrat survey should be carried out.

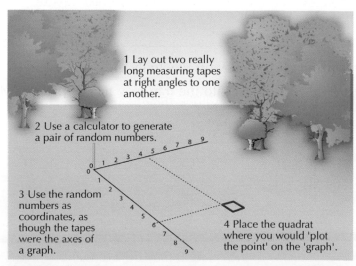

Fig 3.42 *Follow the instructions to ensure a random quadrat survey is carried out*

Pitfall traps

This technique is used to sample organisms living in leaf litter. A **pitfall trap** consists of a jar and a lid. Leaf litter is plant material such as bark, dead leaves, twigs and needles that have fallen on the ground. This is a habitat for many animals, as it is cool and damp and prevents them from drying out.

Using a pitfall trap

A hole is dug in the ground and an empty jam jar or other suitable container sunk into it. The animals living in the leaf litter fall into the trap and are caught. The steep sides make it impossible for them to climb out. The best traps have a lid, which can be made from large leaves or stones (fig 3.43).

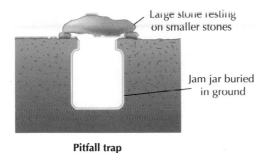

Large stone resting on smaller stones

Jam jar buried in ground

Pitfall trap

Fig 3.43 *The pitfall trap must be level with the soil surface so that animals can fall into the trap*

Limitations and sources of error using pitfall traps

Some limitations and sources of error are noted below.

- The top of the trap is not level with the soil surface. This means that animals may turn around and not fall into the trap.
- Flying animals may be able to fly out of the trap.
- Animals that fall into the trap may not be present when the trap is emptied.
- Animals may die from drowning.

Ways that errors can be reduced

These errors can be reduced in the following ways:

- Dig the hole deep enough and ensure that the top surface of the trap is level with the soil surface.
- Place a lid on top of the trap. This can be done by placing a couple of stones at the edge and resting a large leaf or stone on top. This has to be weighed down by more stones.

- The lid may stop birds from feeding from the trap. Animals that fell into the trap might not be present as they were eaten by carnivores. Empty the trap regularly.

- Place a few small needle holes in the bottom of the trap to allow rainwater to drain away and prevent the animals from dying.

Further sampling techniques

Sampling of organisms can also be carried out using the following methods.

- **Tullgren funnel:** This technique is used to sample soil ecosystems. A Tullgren funnel consists of a funnel with a sieve, a jar to collect animals and a lamp. Tullgren funnels are used to encourage animals living in soil to move out of the soil. Their habitat tends to be cool, damp and dark. A Tullgren funnel makes use of light and heat to drive the organisms out of the soil (fig 3.44).

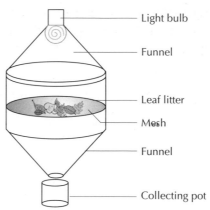

Light bulb

Funnel

Leaf litter

Mesh

Funnel

Collecting pot

Fig 3.44 *A simple Tullgren funnel has only one wire mesh. More can be added to separate animals of different sizes*

Fig 3.45 *The muslin on the end of the sucking tube prevents animals entering the mouth*

- **Pooters:** A pooter allows small animals to be caught. It is a small container with two tubes attached. One end is placed over the animal. The other end is sucked like a straw. It is like an insect hoover. There is no chance of sucking the animals into the mouth, as the sucking end is covered with muslin (fig 3.45). Usually a pooter is used along with tree beating or sweep netting.

- **Tree beating:** This technique is used to sample tree ecosystems. Tree beating is used to dislodge animals from the branches of trees. It requires a metre stick, or branch, white sheet and pooters (fig 3.46).

Fig 3.46 *Usually a white sheet is placed on the ground*

Fig 3.47 *Sweep nets are dragged through the grass in straight lines*

- **Sweep netting:** This sampling method allows animals living in grass ecosystems to be sampled. It requires a sweep net, a pooter and a white tray (fig 3.47). Sweep netting is used to dislodge animals living in long grass.

- **Pond netting:** This technique is used to sample freshwater pond ecosystems. Pond netting sampling allows swimming animals to be caught. It requires a fine mesh net and a white tray (fig 3.48).

Each of the above sampling techniques has its limitations and sources of error. These are summarised in the table below.

Fig 3.48 *Be careful if you try pond netting. Never go into the water. Always take someone with you*

Technique	Limitations and sources of error of the technique	Ways that the errors can be reduced
Tullgren funnel	Animals may die in the soil before they reach the collecting jar.	Place only a thin layer of soil on the sieve.
	Animals may be too large to pass through the mesh in the funnel.	Include several layers of mesh in the funnel. The mesh should have holes of different sizes.
	Carnivorous animals may eat other animals in the collecting dish.	Empty the dish regularly. This will allow more animals to be observed.
Tree beating	Animals may not fall onto the sheet.	Ensure that the white sheet is as large as possible.
	Animals may move off the sheet before they can be caught and identified.	Do not work alone. Lift the sheet by the corners and gently shake all the animals into the middle of the folded sheet.
	Animals may remain on the tree.	Observe the tree being sampled carefully and quietly before beginning sampling. Make a note of flying animals that can be seen on the tree.

Technique	Limitations and sources of error of the technique	Ways that the errors can be reduced
Sweep netting	Animals may not be dislodged from the grass.	When moving through the grass, make sure that the mouth of the net is open. Ensure the net is kept as close to ground level as possible.
	Animals may fly away.	A second sampler could walk behind the first net to catch any flying animals.
Pond netting	Animals may hide around the edge of the pond when they are disturbed by the net.	Sweep the net as close to the edge of the pond as possible.
	Animals may swim away before becoming caught by the net.	Use a figure-of-eight movement when sweeping to catch animals swimming away from the net.
	Animals may swim out of the net as it is being lifted out of the water.	Turn the handle into the net when lifting to close the mouth of the net.
	Animals may be small enough to swim through the holes in the net.	Use a net with as small a mesh possible.

When sampling animals using any of the methods described above, you should release the animals back into their habitat as soon as you have counted them. This helps to avoid ecological damage caused by the sampling technique.

Make the link – Biology

All laboratory-based experiments are repeated to make the results reliable. This should also happen in ecological sampling. Often laboratory-based experiments are repeated quickly after the initial experiment. In ecological sampling the sampling is repeated often on a yearly basis to detect trends.

Make the link – Social studies

You may have learned about weather sampling, which is done to detect trends in climate. What use could ecologists make of the information gathered?

Biology in context

Red deer are the largest wild mammals living in Scotland. They are herbivores. Since the extinction of bears, wolves and lynx in Scotland, adult red deer have no natural predators, although the young are sometimes preyed upon by eagles and foxes. The lack of natural predators allowed the population to increase to a number that the habitat could not support. The red deer reduced plant biodiversity through overgrazing. For this reason, the Forestry Commission monitors the numbers of deer. If numbers rise then a planned cull takes place.

Hint

A *cull* is the removal of animals from a group for specific reasons. Culling usually involves killing animals. Red deer are culled to reduce the damage they cause to the habitat.

GO! Activities

Activity 3.3.1 Working individually

(a) State the steps that need to be taken to ensure representative sampling occurs.

(b) State what is meant by quantitative measurements.

(c) Give the reason why quadrats are used.

(d) Give two limitations or sources of error when using quadrats in sampling.

(e) Describe how the effects of the sources of error or limitations can be minimised.

(f) Explain why pitfall traps are used.

(g) Give two limitations or sources of error when using pitfall traps in sampling.

(h) Describe how the effects of the sources or error or limitations can be minimised.

Activity 3.3.2 Working individually

Quadrats were used to sample the limpets living on a rocky seashore. The quadrats used were 50 cm × 50 cm in size (four quadrats = 1 m²). Three different groups carried out the sampling.

- The positions of the quadrats and the number of limpets found are shown below (fig 3.49).

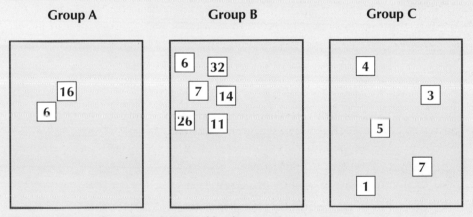

Fig 3.49 *The position of quadrats used by three groups*

(a) Copy and complete the table below to show the estimated number of limpets per m².

Group	Average number of limpets in each quadrat	Estimated number of limpets per m²
A	11	44
B		
C		

(Continued)

(b) State the sampling error made by group 1.

(c) Explain how this error could be minimised.

(d) State the sampling error made by group 2.

(e) Explain how this error could be minimised.

Activity 3.3.3 Working individually

Pitfall traps were carried out by two different groups to sample the animals living in the leaf litter in the grounds of a school.

Both groups left their traps for one day before collecting their results. The results from both groups are shown below.

Group A	Pitfall trap number	Number of each type of animal found in leaf litter				
		Ants	Spiders	Mites	Woodlouse	Earthworm
	1	3	2	1	0	1
	2	2	1	2	0	0

Group B	Pitfall trap number	Number of each type of animal found in leaf litter				
		Ants	Spiders	Mites	Woodlouse	Earthworm
	1	3	1	1	1	0
	2	3	2	1	1	2
	3	0	2	1	1	1
	4	2	0	3	1	1
	5	2	1	2	2	2

(a) State the number of invertebrates found by group A.

(b) State the number of invertebrates found by group B.

(c) Give the number of times more animals found by group B compared to group A.

(d) Explain why group B's results are more reliable than group A's.

(e) State the average number of ants found by group B.

(f) Give one source of error when setting up pitfall traps.

Activity 3.3.4 Working individually

A Tullgren funnel was used to sample the animals living in the leaf litter below an oak tree. The types of invertebrates found in the collecting jar are shown in the table below.

Type of invertebrate	Numbers present in leaf litter
Ants	12
Millipedes	3
Snails	5
Spiders	18
Mites	33
Woodlice	1

The results found were surprising. The experimenter noticed many woodlice and mites in the sample placed into the funnel.

(a) State the percentage of the total number of invertebrates that the spiders represented.

(b) Suggest a reason why there might have been only one woodlouse in the collecting jar.

(c) State how the apparatus could be changed to ensure all the woodlice placed into the funnel were collected.

(d) Suggest what the results indicate about the size of mites.

Activity 3.3.5 Working individually

Tree beating was used to compare the invertebrate animals living on the leaves and branches of an oak tree and a rowan tree. The rowan tree was sampled in February and the oak tree in April.

• The method shown below was used to collect the samples (fig 3.50).

Fig 3.50 *Technique for tree beating*

The samples collected may not have been representative of all the invertebrate animals living on the tree.

(a) Suggest two reasons why the samples may not have included all the animals living in the trees.

(*Continued*)

(b) Suggest a reason why the samples collected could not be compared.

(c) The sampling was repeated for both trees in September so that the results could be compared. Suggest two variables that need to be kept constant so that the sampling gave valid results.

Activity 3.3.6 Working individually

The internet gives you the opportunity to see how sampling can be used to investigate changes in biodiversity across a footpath.

Go to saps.org.uk and click on 'Secondary', then 'Teaching resources' and scroll down to 'Ecology Practical 2—The distribution of species across a path'.

Activity 3.3.7 Working individually

Make your own Tullgren funnel. Do an internet search for a guide to making one. A useful website is ehow.co.uk.

I can:

- State that it is impossible to count every organism present in a habitat because there might be too many organisms to count, it would take too long to count every organism or it may result in habitat damage.

- State that representative sampling requires several samples to be taken.

- State that the larger the area being sampled, the higher the number of samples that need to be taken.

- State that quantitative measurements can be compared, allowing trends to be identified.

- State that a quadrat is used to obtain an estimate of the population in a habitat.

- State that quadrats have to be dropped randomly.

- State that limitations and sources of error when using quadrats include wrongly identified organisms, incorrectly counted organisms or too few samples being taken.

- State that the errors can be reduced by using a key to identify organisms, by making a rule not to include an organism not fully in the quadrat and by taking more samples if the organisms are in clusters.

- State that a pitfall trap is used to trap animals living in the leaf litter.

- State that limitations and sources of error when using pitfall traps include the trap not being level with the soil surface, animals escaping from the trap, animals being eaten while in the trap and animals dying.

- State that the errors can be reduced by ensuring the hole dug is deep enough to house the trap placing a lid on the trap to prevent flying animals escaping, emptying the trap frequently to ensure carnivorous animals don't eat other animals and preventing animals from dying by placing needle holes in the base of the trap.

Abiotic factors

An **abiotic factor** is a non-living condition that affects the growth and distribution of organisms in a habitat. Abiotic factors tend to be related to the weather conditions in the habitat.

Light intensity

Light intensity is an important abiotic factor to sample. Green plants need light to be able to photosynthesise. Plants living in areas with high light intensity are called sun plants. These plants are fast-growing as they receive lots of light.

Shade plants live in areas of low light intensity. This could be in woodland or under the shade of taller sun plants. They are slower-growing plants. Shade plants have adaptations to allow them to survive in areas of low light intensity. Sun and shade plants contain different photosynthetic pigments. Shade plants' leaves tend to be darker green in colour than sun plants.

Light intensity also affects the distribution of animals. Animals living in leaf litter move out of light and into shaded conditions. If they are moved into light they move quickly to reach more shaded conditions. This increases the chance of survival.

A light meter is used to measure the light intensity at the soil surface. It has a light sensor to detect light intensity (fig 3.51).

Fig 3.51 *The light-sensitive layer lies on top of the meter. It is this layer that is pointed towards the area of highest light intensity*

Using light meters

The light-sensitive sensor is pointed at the area of highest light intensity. The needle is allowed to stabilise before the reading is taken.

Limitations and sources of error when using light meters

There are limitations and sources of error in using light meters.

- Light intensity could change during the sampling process. This could be due to cloud cover or sampling at different times of the day.

- The light sensor could be shaded by the users.

Fig 3.52 *A simple thermometer*

Fig 3.53 *The thermometer is left in the soil until the temperature stabilises*

Ways that the errors can be reduced

The following precautions can be taken:

- Aim to take all the readings at the same time of day. This can be difficult. It is better to take several readings at each sample site and work out the average reading.
- Make sure that the light sensor is not pointing towards the users while the reading is being taken.

Temperature

Temperature is an important abiotic factor to sample. The enzymes of each plant and animal have an optimum temperature. Metabolism is controlled by enzymes. Organisms therefore require an optimum temperature to survive.

A soil thermometer or temperature probe can be used to measure the soil temperature. Air temperature can be measured using a thermometer (fig 3.52).

Using soil thermometers or temperature probes

The thermometer (fig 3.53) or probe is gently pushed down into the soil. The thermometer or probe should be left in the soil for a few minutes. When the reading remains constant the reading should be taken.

Limitations and sources of error using soil thermometers or temperature probes

There are limitations and sources of error in measuring temperature.

- The thermometer or probe may not be inserted deeply enough into the soil.
- The reading may not measure soil temperature – it may still be reading the air temperature.
- The bulb of the thermometer is being held in the hand.
- The thermometer may be in direct sunlight.

Ways that the errors can be reduced

The following precautions can be taken:

- Try to push the thermometer or probe into the soil to half its depth. If this is not possible, remove the probe and reposition.
- Allow the reading to stabilise before attempting to take a reading. This will usually need a few minutes.

- It is the expansion or contraction of the liquid in the bulb that indicates the temperature. If you hold the bulb it will measure your body temperature.

- The thermometer must be in shade when measuring air temperature. If it is in direct sunlight the thermometer will give a falsely high reading.

pH

pH is an important abiotic factor to sample. pH is a measure of the acidity or alkalinity of the habitat. Acidic conditions destroy soil nutrients and make it harder for plants to absorb water. Acidic conditions can affect the functioning of fish gills and reduce the oxygen levels in their blood. Some organisms are adapted to live in acidic or alkaline conditions. pH can be measured using a pH probe or pH paper, or by using a chemical test. There are many chemicals that can be used to measure pH.

Using pH meters

The pH meter (fig 3.54) is used in the same way as a moisture meter. There are limitations and sources of error in measuring pH:

- The reading may be contaminated by soil left on the probe from the previous sample.

- Not enough samples might be taken.

Fig 3.54 *A pH meter can test whether soil is acidic or alkaline*

These errors can be reduced by:

- Wiping the probe between each sampling to reduce the risk of cross-contamination.

- Increasing the number of samples that are taken.

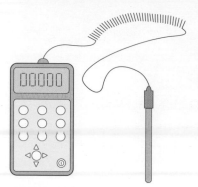

Fig 3.65 *The switch must be switched to moisture when taking a reading*

⚗ Make the link – Biology

In Chapter 19 of this Unit (page **307**), you will learn more about adaptations of animals and plants to living in very dry conditions.

⚗ Make the link – Health and Wellbeing

Think about when you cook. Accurate measurements need to be made to make sure the product turns out as expected. In PE, accurate measurements of pulse rate need to be taken. This allows you to decide if your fitness is improving. Why are accurate readings important when collecting data?

🔍 Hint

A sun plant is a plant that is adapted to live in areas of high light intensity. A shade plant is a plant adapted to live in areas of low light intensity. These plants have different photosynthetic pigments to allow them to absorb light in their habitat.

Soil moisture

Soil moisture content is an important abiotic factor to sample. It measures how much water is in the soil. Most land organisms cannot survive in very wet or very dry conditions. Most plants cannot live in very wet conditions. The water fills up the air spaces in the soil, meaning the plant does not get enough oxygen for respiration. Most plants cannot live in very dry conditions. They wilt and die when water availability is low. Most animals would die of dehydration in very dry conditions. In wet conditions diseases spread more easily.

Using moisture meters

The moisture probe (fig 3.55) is inserted into the soil to a depth of 5 cm. The switch is set to moisture. The needle is allowed to stabilise and then the reading is taken.

Limitations and sources of error using moisture meters

- There may be moisture on the probe from the previous reading.
- Not enough samples are taken.

Ways that the errors can be reduced

The following precautions can be taken:

- Wipe the probe with a paper towel before and after taking each reading.
- Take a repeat for each sample site and calculate an average.

🌳 Biology in context

The Climate Change (Scotland) Act was passed by the Scottish Parliament in 2009. The Scottish Government recognises that 'climate change will have far-reaching effects on Scotland's economy, its people and its environment . . .'

Climate change is determined by measuring abiotic factors over a long period of time. Ecologists measure abiotic factors and pass their findings to Government. The Scottish Government has acted on the ecologists' findings to make plans for the future. Search the Scottish government website to find out more information on plans to tackle the anticipated effects of climate change.

GO! **Activities**

Activity 3.3.8 Working individually

(a) State what is meant by the term 'abiotic factor'.

(b) Name four abiotic factors.

(c) Give the apparatus that would be used to measure the abiotic factors you have named.

(d) State a limitation or source of error that may occur when measuring each abiotic factor.

(e) Suggest a method that could be used to minimise the source of error you have given.

Activity 3.3.9 Working individually

A plant found growing next to burns in Scotland is the marsh marigold (fig 3.56).

Quadrats were used to sample the marsh marigold found growing in 5 different sample sites in the Scottish borders. The quadrats used were 50 cm² × 50 cm² in size (four quadrats = 1 m²).

Fig 3.56 *A marsh marigold*

Two abiotic factors were also sampled and averages were calculated for each area.

The results are shown in the table below.

Site of sampling	Average number of marsh marigold in each quadrat	Estimated number of marsh marigold per m²	Average soil water content (units)	Average soil pH
1	0		10	5.1
2	10		4	6.7
3	15		9	7.3
4	5		3	6.5
5	0		11	5.9

(Continued)

(a) State the estimated number of marsh marigold per m² for each of the five sample sites.

(b) State the ratio of marsh marigolds found at sample sites 2, 3 and 4.

(c) One of the abiotic factors does not appear to influence the numbers of marsh marigold found. Name the factor and explain your reason for choosing this factor.

(d) Soil water content was measured using a light moisture meter. Suggest a source of error that could have occurred when measuring the water content and suggest a way that this error could be minimised.

Activity 3.3.10 Working individually

An investigation into the effect of light intensity on the distribution of a type of plant in a garden was carried out. Eight sample sites were investigated. The light intentisty and the percentage of the ground covered by the plant were recorded. The results are shown in the table below.

Sample site	Light intensity (lux)	Percentage of ground covered by plant (%)
1	1000	80
2	750	70
3	400	15
4	400	15
5	600	20
6	800	40
7	1900	100
8	1200	90

(a) Draw a line graph to show the relationship between sample site and the light intensity **and** the percentage of ground covered by the plant. (Use the same piece of graph paper)

(b) Describe the relationship that exists between the light intensity and the percentage of ground covered by the plant.

(c) Light intensity appears to affect the distribution of the plant. Suggest whether this plant is a sun or shade plant and give a reason to support your suggestion.

I can:

- State that an abiotic factor is a non-living condition that affects the growth and distribution of organisms in a habitat.

- State that four abiotic factors are light intensity, temperature, pH and soil moisture.

- State that to measure light intensity, a light meter is used. To measure temperature, a thermometer is used. To measure pH, a pH meter is used. To measure soil moisture, a moisture meter is used.

- State that an example of a limitation or source of error that may occur when measuring light intensity could be the light sensor being shaded by the users.

- State that an example of a limitation or source of error that may occur when measuring temperature could be the thermometer or probe not being inserted deeply enough into the soil.

- State that an example of a limitation or source of error that may occur when measuring pH could be the probe being contaminated by soil left on the probe from the previous sample.

- State that an example of a limitation or source of error that may occur when measuring soil moisture could be moisture on the probe from the previous reading

- State that a method that could be used to minimise the source of error when measuring light intensity could be to ensure that the light sensor is not pointing towards users.

- State that a method that could be used to minimise the source of error when measuring temperature could be to push the thermometer or probe into the soil to half its depth, allowing the reading to stabilise before taking a reading.

- State that a method that could be used to minimise the source of error when measuring pH could be to wipe the probe between each sampling to reduce the risk of contamination.

- State that a method that could be used to minimise the source of error when measuring soil moisture could be to wipe the probe with a paper towel before and after taking each reading.

Need for keys

When sampling an ecosystem unknown organisms may be collected. It is possible to take the organism to someone who is likely to know. It would be ecologically damaging to the ecosystem if everyone did this. That is why ecologists use keys.

Keys allow easy identification of unknown organisms. They are more reliable than pictures. Organisms show variation. An organism may be collected that does not look identical to the photograph.

Paired statement keys

A paired statement key is a numbered list of statements. Looking at the organism, decide which of the first two statements fits it best. At the end of each statement, a number signifies which statements to use next. The numbers are followed until the name of the organism is found.

Fig 3.57 *Bat*	**Fig 3.58** *Seagull*	**Fig 3.59** *Dolphin*	**Fig 3.60** *Brown trout*	**Fig 3.61** *Bumblebee*

Statement		
1	Animal has wings Animal doesn't have wings	Go to 2 Go to 4
2	Animal has 1 set of wings Animal has more than 1 set of wings	Go to 3 Bumblebee
3	Animal has feathers Animal has fur	Seagull Bat
4	Animal has 2 fins on top of body Animal has 1 fin on top of body	Brown trout Dolphin

The statements are very carefully worded to avoid confusion over characteristics such as colour or size.

Making keys

The diagrams shown below show the invertebrates collected by pupils when sampling an ecosystem. They are not drawn to scale (fig 3.62).

Fig 3.62 *Invertebrates found in a sample of an ecosystem*

The invertebrates can be grouped in several ways:

- Invertebrates with legs – spider, beetle and woodlouse. Invertebrates without legs – earthworm and snail.

- Invertebrates with shells – snail. Invertebrates without a shell – earthworm, spider, beetle and woodlouse.

- Invertebrates with spots on body – beetle. Invertebrates without spots – earthworm, snail, spider and woodlouse.

These groupings can be used to make a paired statement key (see the table at the top of the next page).

Statement 1	
Invertebrates with legs	Go to 2
Invertebrates without legs	Go to 4
Statement 2	
12 legs or more	**Woodlouse**
Fewer than 12 legs	Go to 3
Statement 3	
Spots on body	**Beetle**
No spots on body	**Spider**
Statement 4	
Shell	**Snail**
No shell	**Earthworm**

A paired statement key can be worked backwards to summarise information concerning organisms.

An earthworm 1 'has no shell' – from statement 4

2 'is an invertebrate without legs' – from statement 1

GO! Activities

Activity 3.3.11 Working individually

(a) State why keys are needed when carrying out sampling.

(b) Give a reason why keys are more reliable than pictures at identifying organisms.

(c) State what the numbers at the right-hand side of the key show.

(d) Give the characteristics of a bat using the key found under the heading paired statement key.

(e) State why the statements in a paired statement key have to be carefully worded.

Activity 3.3.12 Working individually

(a) The table below shows some features of five types of white blood cell.

(Continued)

Type of white blood cell	Shape of nucleus	Minimum lifespan of cell	Largest diameter of cell (micrometres)
Neutrophil	Multi-lobed	Hours	12
Eosinophil	Bi-lobed	Days	12
Basophil	Bi-lobed	Hours	15
Lymphocyte	No lobes	Years	15
Macrophage	No lobes	Hours	80

Complete the key using the information given in the table.

1. Nucleus is lobed .... Go to 2
   Nucleus is not lobed .... (1)
2. Minimum lifespan is days .... Eosinophil
   (2) ..... (3)
3. Largest diameter of cell is 12 micrometres ..... Neutrophil
   Largest diameter of cell is 15 micrometres ...... (4)
4. (5) .... Lymphocyte
   Minimum lifespan is hours .... (6)

(b) State three features of a basophil using the key.

Activity 3.3.13 Working individually

The table below shows some features of British ladybirds.

Name of ladybird	Colour of wing cases	Colour of forebody	Number of spots on each wing case
2-spot	Red	Black	1
7-spot	Red	Black	3
10-spot	Red	White	5
Water	Red	Orange	9
18-spot	Brown	Brown	9

(a) Construct a paired statement key from the information in the table. The first statement has been completed but you may wish to construct your own first statement.

1. Red wing case .... Go to 2
   Brown wing case 18-spot

(b) State three features of the water ladybird using the key.

I can:

- State that keys are used to identify organisms.

- State that keys are more reliable at identifying organisms than photographs, as organisms show variations.

- State that a paired statement key is a list of numbered statements.

- State that removing organisms from habitats can cause ecological damage and so can't be used to identify organisms.

19 Adaptation, natural selection and the evolution of species

You should already know:

- Plants and animals show adaptations that allow them to reproduce and survive in their environment.
- Adaptations can be structural, physiological and behavioural.
- DNA influences phenotypes as it codes for proteins.
- Portions of DNA are called genes, and genes code for the production of proteins.
- During cell reproduction, DNA is copied to ensure that each new daughter cell receives an identical copy of the DNA.
- Genetic counselling is offered to prospective parents when potential inherited disease is suspected.

Learning intentions

- Name environmental factors that can increase the rate of mutation.
- Describe what is meant by the term 'mutation'.
- Explain the difference between mutations that are neutral, advantageous or disadvantageous.
- State what is meant by an adaptation.
- Describe adaptations of plants and mammals to living in deserts.
- Explain why organisms need adaptations to live in certain ecological niches.
- Describe what is meant by the term 'natural selection'.
- Explain the processes involved in natural selection.
- Give examples of natural selection in action.
- Describe what is meant by the term 'species'.
- Describe the processes involved in the formation of a new species – speciation.
- Describe examples where speciation in action can be observed.

⚛ Make the link – Biology

Mutations could occur when new cells are being produced. You studied the process of producing new cells in Unit 1 Chapter 3 (pages 38–48). DNA codes for the proteins required by the cell. You studied the role DNA plays in the production of proteins in Unit 1 Chapter 4 (pages 50–56).

Mutation

A **mutation is a random change to the genetic material** of an organism. Mutations are rare, random and **spontaneous**. They cause permanent change to the DNA. The result of a mutation is that the genetic message is changed. This can alter the phenotype or functioning of the organism.

Some mutations are more important than others. Mutations that occur in sex cells are inherited by offspring. They **are the only source of new alleles**. Mutations that occur in body cells can lead to cancer.

There are different types of mutation.

Gene mutations occur when DNA is replicating. Chromosome structure mutations can occur when chromosomes are pairing up during gamete formation. Chromosome number mutations occur during cell division when spindle fibres fail.

Causes of mutations

Mutations occur naturally. They are rare and occur infrequently. A mutation is a fault in DNA. Naturally occurring mutations occur when a cell divides. Prior to cell division, DNA replicates, or copies, itself. If the newly formed copy is not identical to the original DNA, a mutation has occurred.

Certain **environmental factors can increase the rate at which mutations occur**. These environmental factors are called **mutagenic agents**.

Being exposed to mutagenic agents can increase the rate of mutations. Examples of environmental factors that are mutagenic agents are:

- **Radiation** – UV light, X-rays and gamma rays.
- **Some chemicals** – mustard gas, colchicine, caffeine, formaldehyde and chemicals in cigarettes.

Research scientists may expose cells artificially to mutagenic agents. They do this to artificially increase the rate of mutation. It does not cause a specific mutation. By exposing cells to mutagenic agents, scientists hope to find cures for diseases.

Types of Mutations

Mutations conferring an advantage

A mutation can result in an **advantage** for the organism. These are called **advantageous** mutations. They result in an improvement in the organism's phenotype. These mutations are very rare but also very important. They are the source of all new alleles. Advantageous mutations may lead to the evolution of new species as they allow natural selection to occur.

An advantageous mutation can be seen in the peppered moth *Biston betularia*. A random gene mutation produced a melanic moth (fig 3.63). This black moth was camouflaged from predators in industrial areas where trees were no longer covered with lichens and had been blackened with soot. It therefore survived to breed and so passed on its advantageous allele to its offspring. It is considered a classic example of natural selection in action.

The failure of all spindle fibres during cell division in plants can be an advantageous mutation. The resulting plant is called a polyploid. They are known to possess increased resistance to

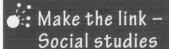

Make the link – Social studies

'Agent Orange' is a powerful herbicide sprayed by Americans on vegetation during the Vietnam War. It was hoped that the enemy would be more visible when the trees dropped their leaves. It was then discovered to be a powerful mutagenic agent. No one realised the birth defects that it would cause. Can you think of any other mutagenic agents that have been used during warfare?

Hint

Gametes are sex cells. If a mutation occurs during gamete production new alleles may be formed. Alleles are the variations that exist for one particular gene. This new allele may be expressed in the organism's phenotype. The phenotype is the physical appearance of the organism.

Fig 3.63 *Peppered moth*

Fig 3.64 *Cystic fibrosis is an example of a disadvantageous mutation. Sufferers have difficulty breathing due to the thick, sticky mucus they produce*

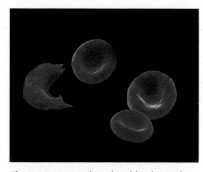

Fig 3.65 *Normal and sickle-shaped red blood cells*

disease. They are larger in size and grow more vigorously than their non-polyploid relatives. Many cereal plants are polyploid. This is important economically.

Mutations conferring a disadvantage

A mutation can result in a **disadvantage** for the organism. These are called **disadvantageous** mutations. If the mutation occurs in genes that code for vital enzymes the organism may not survive. Chromosome structure mutations where genes are lost tend to be lethal. Disadvantageous mutations may lead to extinction.

A disadvantageous gene mutation causes cystic fibrosis. Cystic fibrosis is one of the UK's most commonly inherited diseases. A mutation in the gene that controls the movement of salt and water in and out of the cells within the body causes cystic fibrosis. The condition affects the lungs and digestive system, which become clogged with thick mucus. This results in difficulty breathing and digesting food (fig 3.64). More than 9,000 people in the UK have cystic fibrosis.

A disadvantageous gene mutation causes sickle-cell anaemia. People receive two alleles for each characteristic. People with one mutated allele are normally healthy. They may even have a selective advantage over people with two normal alleles in areas where malaria is present. People with two sickle cell alleles have the disease and the disadvantage. Their blood tends to form clots and they are unable to carry as much oxygen. Their red blood cells have a characteristic sickle shape (fig 3.65).

Neutral mutations

Mutations with no effect on the organism are called **neutral** mutations. These mutations occur in part of the DNA that does not code for a protein. They can also occur in a gene that is not expressed in the phenotype. An incorrect amino acid can be incorporated into the protein but it does alter the function of the protein. An extra toe would be a neutral mutation (fig 3.66).

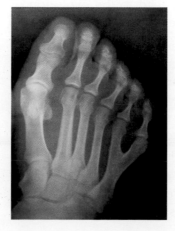

Fig 3.66 *This person has an extra toe. There is neither an advantage nor a disadvantage to this. It is a neutral mutation*

GO! Activities

Activity 3.4.1 Working individually

(a) Give a definition for the term 'mutation'.

(b) State three possible effects on an organism that a mutation may have.

(c) Give two words that could be used to describe mutations.

(d) State two environmental factors that could increase the rate of mutations.

(e) Give the term that is used to describe environmental factors that bring about mutations.

Activity 3.4.2 Working individually

Mutations can be caused by being exposed to radiation and from chemicals. A study was carried out into the sources of radiation and how much each source contributed to the total exposure.

It was found that natural radioactivity in the air accounted for 38% of the total radiation. The ground and buildings exposed people to exactly half of the natural radioactivity in the air. Surprisingly, 16% of the radiation came from food and drink. Medical X-rays accounted for 13%. Cosmic rays from space accounted for 1% more than medical X-rays.

(a) Present the information given above as a table.

(b) State the percentage contribution that comes from natural radioactivity and medical X-rays.

Activity 3.4.3 Working in pairs

Cigarette smoke is known to contain chemicals that are mutagenic agents.

(a) On the internet, go to smokefree.nhs.uk and click on 'Advice and Information' and then 'Behind the Campaign'.

Your task is to produce an information leaflet that could be given to smokers to explain the risk of developing cancer as a result of smoking.

You should:

- State the number of mutations needed for a normal cell to turn into a cancer cell.
- Explain why most types of cancer are more common in older people.
- Explain why smoking could lead to cancer.
- State the number of cancers that are caused by smoking.
- Explain why heavy smokers are more likely to develop cancer.
- Include a photograph of healthy lungs, and lungs that have been damaged by smoking.

Your leaflet should be one folded A4 page.

I can:

- State that a mutation is a random change to the genetic material of an organism.
- State that mutations are rare, random and spontaneous.
- State that mutations are the only source of new alleles.
- State that certain environmental factors can increase the rate at which mutations occur.
- State that these environmental factors are called mutagenic agents.
- State that radiation and some chemicals are mutagenic agents.
- State that X-rays, UV light and gamma rays are examples of types of radiation that cause mutations.
- State that mustard gas, colchicine, caffeine, formaldehyde and chemicals in cigarettes are examples of chemicals that cause mutations.
- State that mutagenic agents do not cause specific mutations.
- State that a mutagenic agent can be used to artificially speed up the rate of mutations.
- State that mutations that are neutral cause no advantage or disadvantage to the organism.
- State that disadvantageous mutations cause harm to organisms.
- State that cystic fibrosis and sickle-cell anaemia are caused by disadvantageous mutations.
- State that advantageous mutations are the source of new alleles in a species.

Variation

All members of a **population** show **variation.** A variation is simply the differences between members of the population. Variation results from differences in genetic information the organisms contain. In ideal conditions, all members of the population could survive until old age. In reality, most organisms fail to survive to reproductive age. The organisms that do survive allow a population to **evolve over a period of**

time in response to changing environmental conditions. This is because they are able to pass on the beneficial variations to their offspring.

Adaptations

Certain variations make it more likely that the organism will survive. These variations are called **adaptations**. An adaptation is an **inherited characteristic that makes an organism well suited to survive in its niche**. Organisms that are well adapted cope better with changes that occur in their environment. They are better at competing for food, escaping from predators and fending off disease. They are more likely to survive and produce offspring. The alleles that allowed them to survive will be passed on to the offspring.

Make the link – Biology

You learned how alleles are passed on to offspring in Unit 2 Chapter 13 page 156. You should be able to work out how even a recessive allele gradually changes the population if it is advantageous.

Desert adaptations

Special adaptions are needed for survival in extreme habitats. Deserts are hot and dry, and there may be very little rain. There is very little shade, and daytime temperatures are very hot. At night it is very cold.

Desert rat

The desert rat is a good example of a small mammal that shows both behavioural and physiological adaptations (fig 3.67).

Fig 3.67 *Desert rats are nocturnal mammals*

- Behavioural adaptations – the way the desert rat alters its behaviour to conserve water. Desert rats remain in underground burrows during the day and are active at night – they are nocturnal.

- Physiological adaptations – the way the desert rat's body functions is designed to conserve water. These animals do not sweat and produce very concentrated urine.

Camels

Camels are a good example of large mammals that show structural adaptations (fig 3.68).

Fig 3.68 *Camels are well adapted to living in hot and dry conditions*

- Structural adaptations – physical features of camels that allow them to survive in harsh conditions. These animals have large feet so that their weight is evenly spread, preventing them sinking into the sand. Their nostrils are slits and they have a double layer of eyelashes to keep sand out. Their fur is very thick on top of their body to protect their skin from the sun but thinner elsewhere to allow heat to be lost.

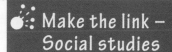 Make the link – Social studies

You may have studied deserts in detail. You will know that the conditions are harsh and that water is not readily available. What other requirement needed by plants may be missing from desert soil?

 Hint

Transpiration is the evaporation of water from leaf surfaces. Water is lost through the stomata.

Desert plants

Cacti are good examples of plants that are adapted to live in deserts (fig 3.69).

- Structural adaptations – they have stems that store water. Instead of leaves they have spines to reduce water loss by transpiration. They have widespread root systems allowing them to obtain water from a large area.

Fig 3.69 *Desert plants are well adapted to living in arid conditions*

Changing environmental conditions

An organism's niche can change due to environmental conditions. These could include natural disasters such as drought, floods and fires. As a result the habitat changes. These changes to the habitat could alter the population in several ways.

Animals – Over time the quantity of food, water and availability of shelter may change. Animals also have to survive changes to diseases and predators in their environment.

Plants – Over time the quantity of soil nutrients, light and water may change in the environment. Plants also have to survive changes to diseases and herbivores in the environment.

Organisms that have adaptations allowing them to survive will pass on these adaptations to their offspring.

🌳 Biology in context

Humans are not well adapted to living in desert conditions. Each year people die in Death Valley national park in the US. There have been recent examples of satellite navigation taking people on an incorrect route. They run out of water and are not adapted to living in the harsh conditions.

GO! Activities

Activity 3.4.4 Working individually

(a) State why it is important that organisms show variation.

(b) State what is meant by the term 'adaptation'.

(c) Explain why organisms living in deserts need to be adapted to survive.

(d) State two factors that might cause environmental conditions to change.

(e) State the three broad categories to which all adaptations can be classified as belonging.

Activity 3.4.5 Working individually

The diameter of a barrel cactus was measured before and after rainfall (fig 3.70). The measurement was found to be
32 cm before the rain fell and 44 cm after rainfall.

(a) State the percentage increase in the diameter of the cactus after rainfall.

(b) Suggest an adaptation that this type of cactus has to allow it to increase in diameter.

(c) The leaves of these cacti are reduced to spines to reduce water loss. Suggest another advantage of having spines instead of leaves.

Fig 3.70 *A clump of barrel cacti*

Activity 3.4.6 Working in pairs

The photograph below shows a moth pollinating a flower at night (fig 3.71)
Think about the adaptations that the flower shows.

(a) The flower is white. Explain how this adaptation helps the moth to act as the pollinator.

(b) The flower usually produces a strong and sweet-smelling scent. Explain how this adaptation helps the moth to act as the pollinator.

(c) The flower has a long nectar tube. Explain how this adaptation helps the moth to act as the pollinator.

Fig 3.71 *Moths can transfer pollen from flower to flower*

I can:

- State that all members of a population show variation.
- State that variation is the differences between members of the population.
- State that the variations allow organisms to evolve over time in response to changing environmental conditions.
- State that variations that make it more likely that the organism will survive are called adaptations.
- State that an adaptation is an inherited characteristic.
- State that special adaptations are needed for survival in extreme habitats such as deserts.
- State that adaptations can be behavioural, physiological and structural.
- State that an example of a behavioural adaptation is the desert rat living in an underground burrow during the day.
- State that an example of a physiological adaptation is the desert rat, which does not sweat.
- State that a structural adaptation is seen in camels, which have large feet to spread their weight over the sand.

Natural selection

Fig 3.72 *Charles Darwin proposed the theory of evolution*

All organisms produce offspring to carry on their species. **Species produce more offspring than the environment can sustain**. This leads to a 'weeding out' process. **Natural selection, or survival of the fittest, occurs when there are selection pressures**. Selection pressures are different for different individuals of the species. Those not adapted may die. **The best adapted individuals survive to reproduce** and **pass on the favourable alleles that confer the selective advantage** to their offspring.

The idea of natural selection was first put forward by Charles Darwin and Alfred Wallace in 1858. Darwin published his book 'On the Origin of Species' where he explained his concept of natural selection (fig 3.72).

Darwin had been employed as a naturalist on the HMS Beagle. He had spent many years studying organisms and had noted patterns. His studies allowed him to conclude:

1. Species produce many more offspring than the environment can sustain.

2. All offspring show variations. Some variations many be advantageous, and others may be disadvantageous.

3. This leads to a struggle for survival. For example they struggle to obtain food and mates, to avoid predation and disease. These are the selection pressures.

4. Only the best-adapted organisms with the favourable alleles are able to survive to reproductive age.

5. These organisms pass on the alleles that confer the selective advantage.

6. The offspring produced experience selection pressures.

7. Only those that are best adapted survive to reproductive age.

8. Gradually the whole population changes to possess the alleles that confer the selective advantage.

Through these stages, Darwin was able to explain the evolution of organisms best adapted to their environment. It is important to remember that characteristics of a population change gradually from generation to generation. Only the fittest individuals survive to pass on the favourable alleles. The process takes time.

Natural selection in action

People may believe that once an organism has adapted to its environment there is no need to change further. This is not the case. The environment is constantly changing so organisms have to constantly evolve. Natural selection continues to change the characteristics of a population. It does this over a long period of time. There are some examples where the process of natural selection takes place fast enough to be observed.

Peppered Moth

A classic example is that of the peppered moth *Biston betularia*. Two forms of the moth exist. The light form is dominant, and the dark one arose from a mutation (fig 3.73). The light one possessed a selective advantage, as it was able to camouflage

Fig 3.73 *Two forms of the peppered moth exist*

Fig 3.74 *Light moths possess a selective advantage on lichen-covered trees. Dark moths possess a selective advantage on trees where lichens have died*

against lichen-covered tree trunks (fig 3.74). The dark moth had a selective disadvantage, as it was easily spotted by predators.

During the industrial revolution in 19th-century Britain sooty deposits killed lichens, and surfaces became encrusted with soot. The dark moth was now camouflaged. It had the selective advantage. It survived to reproduce and pass on the dark allele (Fig 3.73). Gradually the population changed in industrial areas so that more dark moths appeared.

The 'Clean Air Act 1956' reduced pollution released by factories. The lichens on trees in industrial areas recovered. Gradually the number of dark moths decreased as they were easily seen by predators and eaten.

Resistant bacteria

Bacterial infections have been commonly treated with antibiotics since the 1950s. This has led to the development of resistant strains of bacteria. Natural variations in the populations of bacteria meant that some were not killed by the antibiotic. The sensitive bacteria were killed, leaving the antibiotic-resistant bacteria to thrive. They had a selective advantage. New types of antibiotics were produced, but over time all have become less effective.

🌳 Biology in context

MRSA is a type of bacterial infection. It is often carried in the nostrils and throat, and on the skin. It is difficult to treat. Through natural selection it has become resistant to many widely used antibiotics. People in hospital are most at risk. This could be because their immune systems are weaker. Also, after an operation there may be an open wound that could become infected. Good hygiene has reduced the number of cases in recent years.

GO! Activities

Activity 3.4.7 Working individually

(a) State why organisms produce offspring.

(b) State which members of a population survive to reproduce. Explain why this is advantageous.

(c) Give two resources that organisms struggle to obtain.

(d) Explain why the melanic form of peppered moth possessed a selective advantage during the Industrial Revolution in British cities.

(e) Suggest why the light form of peppered moth possessed a selective advantage in northwest Scotland, even during the Industrial Revolution.

Activity 3.4.8 Working in groups

Darwin's finches are found on the Galapagos Islands. They are thought to have evolved from one type of ancestral finch blown over from the mainland (fig 3.75).

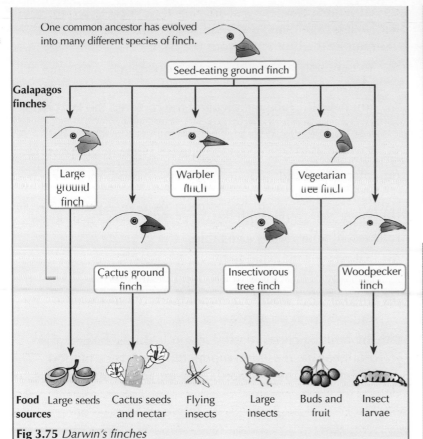

Fig 3.75 *Darwin's finches*

(a) State two quantitative measurements that could be taken of the birds' beaks.

(b) From the diagram, give one environmental factor that could have led to this variation.

(c) Suggest which finch is most similar to the ancestral finch. Explain your choice.

(d) Suggest which finch is most dissimilar to the ancestral finch. Explain your choice.

Activity 3.4.9 Working in groups

To do this activity you will need:

- Bulldog clips of different sizes
- A Petri dish for each group member
- A small food container filled with broth mix
- A stopclock.

1. Each member of the group should select a bulldog clip of a different size.

2. Each member has a Petri dish on the desk beside them.

3. Place the food container filled with broth mix in the middle of the group.

Make the link – Biology

In Unit 2 Chapter 13 (pages 156–173) you learned about dominant and recessive alleles. Mutations are the source of new alleles. Could you explain why a recessive allele may not show up in the phenotype of an affected individual?

Make the link – Religious and moral education

You have learned the theory of natural selection. This is not the view taken by many religions. Some people believe that life on Earth exists exactly as God created it. Could you put forward an argument to support or refute the theory of natural selection?

Hint

Remember that natural selection is still continuing in organisms. New variations occur due to mutation. If the mutation has a selective advantage, the whole population will gradually evolve to possess the mutant allele. This takes a very long time to happen.

(Continued)

4. Each group member should use the bulldog clip to remove the broth mix from the dish.

5. You are only allowed to remove one piece of the mix at a time.

6. You have to place the piece of the mix into the Petri dish.

7. The time allowed is 30 seconds.

8. Your aim is to collect as many pieces of the mix as possible.

(a) After 30 seconds, count the pieces of broth mix of each type you have.

(b) Share your results with the other members of your group.

(c) Present your results as a table.

(d) State the relationship between the size of bulldog clip and size of the piece of broth mix collected.

(e) Suggest what would happen if the broth mix was replaced with dried peas.

(f) The bulldog clips are used to model birds' beaks. The broth has been used to model different types of food. Explain how these results apply to natural selection.

I can:

- State that all organisms produce offspring to carry on their species.

- State that organisms produce more offspring than the environment can sustain.

- State that variations exist in offspring produced.

- State that survival of the fittest is based on the idea of a weeding-out process.

- State that only the best-adapted individuals survive to reproduce.

- State that alleles that confer a selective advantage are passed on to offspring

- State that natural selection is a continuing process.

- State that natural selection requires a long time to adapt populations to their habitats.

- State that natural selection in action was seen in peppered moths.

- State that the dark moth arose due to a mutation.

- State that the dark moth had a selective advantage in industrial areas and avoided predation.

- State that the proportion of dark moths in the population gradually increased as the dark moths were able to pass on the selective advantage to their offspring.

Species

Organisms that are able to interbreed and produce fertile offspring are called a species. Being fertile means that the offspring are able to reproduce and produce offspring themselves.

Some closely related organisms are able to interbreed, but the offspring are sterile. Horses and donkeys are closely related. They are able to interbreed and produce offspring but the offspring are infertile and cannot reproduce. A cross between a male horse and a female donkey produces a mule (fig 3.76). Lions and tigers are also closely related and produce offspring called ligers. The offspring are infertile (fig 3.77).

Organisms that are not closely related are unable to produce offspring with each other.

Fig 3.76 *A mule is sterile and unable to produce offspring*

Process of speciation

Speciation is a process that allows new species to develop. Members of the population become isolated when populations of the same species follow different evolutionary paths due to natural selection. The organisms with the genes best suited to the selection pressures survive to breed.

Speciation requires several processes to take place.

Fig 3.77 *A liger is unable to produce offspring, as it is sterile*

1. **Isolation**

   Members of the original **population become isolated**. This prevents the members of the new subpopulations from interbreeding with each other and the original population. The exchange of genes by interbreeding is prevented. The isolation can be brought about by:

   - Geographical barriers – these could be seas, rivers, mountains or deserts.

   - Reproductive barriers – such as sex cells that cannot fuse, mating displays that don't attract members of another population or flowering at different times of the year.

   - Ecological barriers – resulting from variations in temperature, water availability and pH in different ecosystems.

2. **Mutation**

   In each of the isolated **subpopulations mutations occur**. Remember mutations are spontaneous, rare and random. Therefore the mutations that occur on either side of the isolating barrier will be different. Mutations create new alleles in each of the populations. These alleles did not exist in the original population.

3. **Natural selection**

   Natural selection selects for different mutations in each subpopulation. Each subpopulation will be exposed to **different selection pressures**. These could be due to differences in climate, predation or diseases. If the mutation gives members of the subpopulation a selective advantage it will be selected for. Those organisms not possessing the advantageous mutation will be selected against. The beneficial mutation will be passed on to offspring.

4. **Time**

   After many generations, mutations and natural selection drastically change the gene pool of each population. They will become distinct groups. If the barrier is removed they will no longer be able to interbreed and produce fertile offspring. **Therefore, each subpopulation evolves until they become so genetically different they are two species.**

Rapid selection

MRSA

Bacteria such as MRSA have become resistant to antibiotics due to natural selection. Each bacterium shows variation. Some will have natural resistance to antibiotics. The antibiotic kills the sensitive bacteria. The remaining bacteria have a selective advantage and survive to pass on the resistance to their offspring. If these bacteria are exposed to the same antibiotic again they survive and multiply. Eventually, a population of bacteria evolves that is completely unaffected by the antibiotic. Since some bacteria can reproduce every 20 minutes, the whole population can become resistant in a very short period of time (fig 3.78).

GM crops

GM crop plants have been genetically modified to lessen the effects of intensive farming on the environment. To be economical, farmers want to produce the highest yield possible. They spray insecticides to stop the crop being eaten. The insecticide costs money. GM crops have been genetically engineered to contain a gene that produces a toxin (fig 3.79). Insects die if they eat the crop. Fewer chemicals are used due to inbuilt resistance of the plant.

Since this technique has been in use, the effectiveness has decreased. This is because some insects had a natural resistance to the toxin produced by the plant. They survived and passed on the advantageous allele to their offspring. This resulted in a greater loss of crops (fig 3.79).

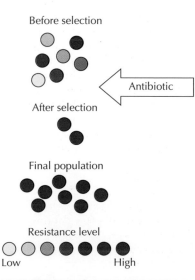

Fig 3.78 *Natural variation means that some bacteria have resistance before the antibiotic is used. After the antibiotic is used, all bacteria are resistant*

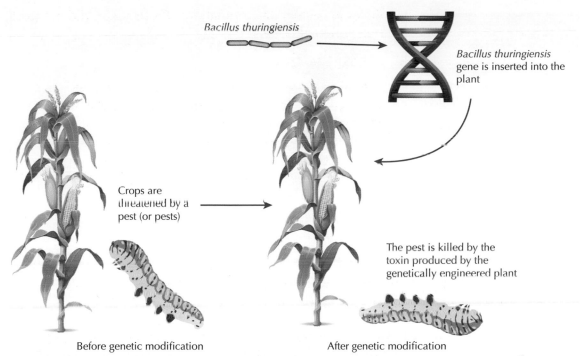

Bacillus thuringiensis

Bacillus thuringiensis gene is inserted into the plant

Crops are threatened by a pest (or pests)

The pest is killed by the toxin produced by the genetically engineered plant

Before genetic modification

After genetic modification

Fig 3.79 *GM crops are engineered to contain a gene that kills insect pests. Some resistant insects are not killed. The whole population rapidly changes to contain the resistant allele*

GO! Activities

Activity 3.4.10 Working individually

(a) Explain what is meant by the term 'species'.

(b) Give the term that is used to describe the process that allows a new species to develop.

(c) Explain why isolation is needed in the formation of a new species.

(d) Explain why mutation is needed in the formation of a new species.

(e) Explain why natural selection is needed in the formation of a new species.

Activity 3.4.11 Working individually

Isolation is important in the process of speciation. Members of the same species share a common gene pool. The frequency of the alleles of genes is maintained by random mating. During the process of speciation, members of the same species are prevented from interbreeding by isolation barriers. The diagram below shows the land mass on Earth (fig 3.80).

♣ Biology in context

On the island of Arran, two of Scotland's rarest tree species are found. The trees do not exist anywhere else in the world and there are only thought to be a couple of hundred of them. These trees are the Arran whitebeam (first observed in 1897) and the Arran cut-leaved whitebeam (first observed in 1952). The Arran cut-leaved whitebeam is thought to have arisen from a cross between the Arran whitebeam and a rowan tree. They have arisen due to isolation leading to speciation. As their habitat is reduced so the number of these rare trees has declined. WWF classes these trees as dangerously near to extinction.

(Continued)

Make the link –Biology

Speciation has led to insect resistance to toxins produced by GM crops. You have already learned about genetic engineering (Unit 1 Chapter 6 pages 70–74) and should know how the insect toxin was introduced into the crop plant.

Make the link – Social studies

Archaeologists look for clues to events that happened in the past by carrying out digs to look for ancient relics. How do biologists gain evidence to support the theory of evolution?

Hint

Be very careful when naming the different types of isolation barriers. People often make mistakes. You should remember **geographical** and **ecological**.

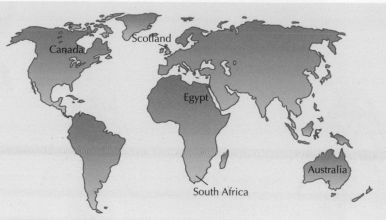

Fig 3.80 *Land mass on Earth*

Isolation barriers can be geographical barriers, reproductive barriers or ecological barriers. Geographical barriers include seas, rivers, mountains and deserts. Ecological barriers result from variations in temperature, water availability and pH of different ecosystems. Reproductive barriers include sex cells that cannot fuse, mating displays that don't attract members of another population or flowering at different times of the year.

(a) Give three types of isolating barriers.

(b) Suggest the type of barrier that could prevent a plant living in Canada interbreeding with a plant in Scotland. Explain your choice.

(c) Suggest the type of barrier that could prevent a plant living in South Africa interbreeding with a plant living in Egypt. Explain your choice.

(d) Suggest the type of barrier that could prevent a plant living in northern Australia interbreeding with a plant in southern Australia. Explain your choice.

Activity 3.4.12 Working in pairs

Go to www.kilda.org.uk and follow the link to the 'Wee Kilda Guide'. From there, find out about the St Kilda Wren.

St Kilda is an island found in the North Atlantic Ocean. It is over 100 miles away from the Scottish mainland. The St Kilda wren is found there. Research the ways in which the St Kilda wren is different from the mainland wren.

Include:

- Differences in overall size, wing length, bill thickness and leg thickness.

- Appearance, including colouring and markings.

- Size and weight of eggs.

- The natural habitat of the birds.

    (a) Suggest reasons for these differences.

    (b) Predict what might happen if the St Kilda wren was introduced to the Scottish mainland.

Activity 3.4.13 Working in groups

The theory of evolution suggests that all life has evolved from a common origin. Over millions of years and as a result of many occurrences of speciation, life on Earth has become varied to such an extent that it would be easy to argue that a common origin did not exist.

The 'Tree of Life' shows the interconnectedness of all life.

    (a) Search for 'Open University tree of life' and follow the link to get to the interactive tree of life diagram.

Use the information in this website to decide whether you agree or disagree with the theory of evolution.

I can:

- State that a species is a group of organisms that can interbreed and produce fertile offspring.

- State that speciation is the name given to the process where two new species are formed from one original species.

- State that the process of speciation requires isolation, mutation, natural selection and time to occur.

- State that isolation is needed in speciation, as it prevents the members of the new subpopulations from interbreeding with each other and the original population.

- State that the three isolation methods are geographical, ecological and reproductive isolation.

- State that mutation is needed to create new alleles in each of the populations.

- State that the mutations in each subpopulation are different.

- State that these alleles did not exist in the original population.

- State that natural selection is needed in speciation, if the mutation gives members of the subpopulation a selective advantage it will be selected for.

- State that the organisms with the selective advantage will survive to pass on the beneficial allele to the offspring.

20 Human impact on the environment

Increasing human population

The total human population has grown significantly since the 1950s (fig 3.81). The massive increase in population is called a population explosion. The main increase has occurred in developing countries. Increased life expectancy resulting from better medicines and healthcare account for the increase in world population. Unfortunately many people have died due to famine – lack of food. Developed countries have a more stable population.

The world population is expected to almost double by 2050. The **increasing human population requires increased food yield**. Therefore it is necessary to change and improve farming

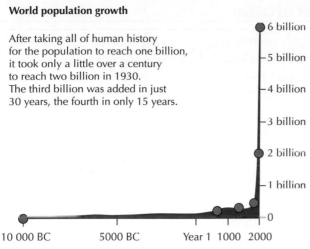

World population growth

After taking all of human history
for the population to reach one billion,
it took only a little over a century
to reach two billion in 1930.
The third billion was added in just
30 years, the fourth in only 15 years.

6 billion
5 billion
4 billion
3 billion
2 billion
1 billion
0

10 000 BC   5000 BC   Year 1   1000   2000

Fig 3.81 *The human population remained fairly stable until recently. The dramatic increase in growth is called a population explosion*

methods to produce food for the expanding population. It is hoped that this will be achieved by increasing intensive farming practices including switching to monocultures.

Intensive farming

Intensive farming is used to increase food production. Farmers grow crops known to produce a high yield. There is a heavy reliance on chemical fertilisers and pesticides.

Features of intensive farming methods

- Competing plants are removed

  Soil is ploughed before planting. Unwanted plants that grow from naturally dispersed seeds are killed using herbicides. This allows more of the nutrients added to the soil to be taken up by the crop. However herbicides kill plants and so reduce biodiversity.

- Herbivorous animals are removed

  Pesticides are sprayed onto the crop, killing animals that would eat the crop. The plants are less damaged and photosynthesise more. More energy can be transferred to the intended consumer.

- Animals are kept inside

  This is called 'battery farming'. Animals live in small spaces, as seen in the battery farming of chickens (fig 3.82) or fish reared in fish farms. They lose less energy in movement so more energy can be transferred to the intended consumer. Rearing animals in this way allows growing conditions to be closely monitored. However living in small areas makes it easier for disease to spread.

Fig 3.82 *Battery-farmed chickens receive exactly what they need to grow. Disease can spread more quickly*

🔍 Hint

Remember that monocultures are an example of intensive farming. They are not the only way that food is produced intensively. Monocultures use a lot of chemicals to maintain crop yield.

Fig 3.83 *Field sizes were small in the past. The land was used to grow the different crops required to feed both the people and the animals that lived on the farm*

Fig 3.84 *A huge area of land is devoted to growing the single plant type*

Fig 3.85 *Heavy farming machinery compacts the soil and can lead to a decrease in soil quality*

☀: Make the link – Biology

In Unit 3 Chapter 19 (pages **316–317**), you learned that GM crops were designed to increase the yield of food produced. There are only two GM crop plants authorised for cultivation in the EU. One is an insect-resistant maize plant and the other a potato with modified starch content. Neither of these types of plants is suitable for growing in Scotland.

Monoculture

Traditionally in Britain farming was done on a small scale. People grew a variety of crops. The crops they grew provided the food for them and their animals. Anything left over was sold. The problem was that there was never much left over (fig 3.83). As the population grew, farming methods changed. Farmers switched to monocultures to provide food for the increasing population.

Monocultures are an example of intensive farming of plants. It is a type of farming where only one variety of crop is grown. The plants are normally genetically identical. Huge areas of land are used to grow the single crop (fig 3.84). Monocultures rely heavily on chemicals such as pesticides and fertilisers.

Advantages of monocultures

- The crop grown is selected as it is suitable for the soil type and climatic conditions.

- More machinery can be used by increasing the area used to grow a single crop. This results in lower labour costs and speeds up planting and harvesting.

- Some monocultures may be able to support more than one harvest per year.

All these features contribute to increased crop yield.

Disadvantages of monocultures

- As the crops are genetically identical the whole crop is equally at risk of disease. Disease also spreads more rapidly, since the plants are grown close together. Herbivores may move into the area to feed on the plentiful supply of food. As a result of an increase in the risk of disease and increased pest numbers, monocultures use more pesticides. This reduces biodiversity.

- The same crop is grown year after year on the same land. The nutrients that the crop requires are used up, and the soil doesn't get a chance to recover. Monocultures require a lot of fertilisers to maintain yields. This can lead to an increase in algal blooms.

- The heavy machinery used in monocultures can compact the soil. This can reduce the oxygen available to the crop root systems for respiration. This can result in a decrease in crop yield (fig 3.85).

- To increase field size, hedgerows are removed. This reduces the habitat available to wild animals. Biodiversity is reduced as a result.

Biology in context

Many countries have a history of growing 'cash crops'. These are crops that are grown for their high monetary value. The problem is that they tend to be grown in countries where people do not have enough food to eat. Valuable land that could be used to feed the population is turned over to growing crops such as coffee and tea, which can be sold abroad for money.

 Make the link — Social studies

The Industrial Revolution changed life for many people in Britain. People moved off the land and into cities. This resulted in the formation of transport links. What effects do transport links have on the environment?

GO! Activities

Activity 3.5.1 Working individually

(a) State why the human population needs to increase the production of food.

(b) Explain why intensive farming is used to produce food.

(c) Explain why intensively farmed animals are prevented from moving around freely.

(d) Give three advantages and three disadvantages of monocultures.

(e) Explain why there is a high use of fertilisers in monocultures.

Activity 3.5.2 Working individually

The information contained in the passage below is adapted from the following website: http://www.forestry.gov.uk/forestry/INFD-7TKF5F

Many people feel that monocultures reduce biodiversity. The Forestry Commission Scotland (FCS) puts forward an alternative point of view when discussing forest in Dunoon. The FCS admits that a type of spruce tree called Sitka when grown in monocultures 'can drain the soil of nutrients and crowd out other plants' when the tree is densely planted.

However, the FCS argues that when the spruce is planted as a monoculture, the trees form a dense canopy that prevents light reaching the forest floor. This provides a suitable habitat for many animals such as deer and foxes. The dense tree trunks provide them with cover and provide a sheltered habitat. The canopy also provides a habitat for many bird species such as goldcrests and siskins. These birds are the food source for birds of prey, which feed from the canopy on these smaller birds.

Gales coming from the Atlantic are strong enough to bring down some trees. The FCS states that when this happens, the trees are left uprooted to provide insects with a habitat. This is beneficial as small birds feed on the insects. When the land is harvested, and the trees are felled, many dormant seeds begin to grow due to the increase in soil temperature. Plants that grow when the spruce trees are felled include mosses, ferns and heather. Some of these seeds may have been dormant for up to 20 years in the leaf litter.

(a) State a disadvantage mentioned in the passage of growing spruce trees in a monoculture.

(Continued)

(b) Give two examples of mammals that obtain a sheltered habitat from spruce trees.

(c) Name two birds that can be found living in the canopy.

(d) Suggest why other plants are unable to grow on the forest floor.

(e) Suggest why these forests may have a high biodiversity of birds of prey.

(f) Suggest why birds such as goldcrests may move into the habitat shortly after a storm.

(g) Name three plants that may lie dormant in the leaf litter.

(h) Suggest which abiotic factor prevents these plants from germinating before the spruce trees are felled.

(i) Predict the effect of felling the spruce trees on plant biodiversity.

Activity 3.5.3 Working individually

Every year, millions of intensively farmed chickens die before reaching the human food chain due to poor living conditions. Chickens are intensively farmed to reduce the cost of food to the consumer.

Many birds are kept in barns and may only have an area the size of an A4 piece of paper in which to live. They are exposed to constant dim light. This encourages them to feed continuously and discourages them from moving around. Wastes such as ammonia build up in the barn. This can cause leg burns, as ammonia is acidic.

In 2012, new laws were introduced to improve the living conditions of intensively farmed chickens in the European Union.

(a) Suggest why farmers want to prevent the birds from moving around in the barn.

(b) Suggest why farmers want the birds to feed continuously.

(c) Suggest the source of ammonia in the barn.

I can:

* State that the biggest increase in human population is occurring in developing countries.

* State that, compared to developing countries, developed countries have a more stable population growth.

* State that a sudden increase in population numbers is called a population explosion.

* State that intensive farming techniques increase food yield.

* State that intensive farming techniques rely heavily on the use of chemicals.

- State that intensive farming allows more energy to be transferred to the intended consumer, since pests are removed.

- State that an advantage of using monocultures is that the crop grown is suitable for the climatic conditions.

- State that an advantage of using monocultures is that more machinery can be used so labour costs are reduced. More machinery means planting and harvesting are speeded up

- State that an advantage of using monocultures is that more than one harvest per year can be supported.

- State that a disadvantage of using monocultures is that crops are genetically identical so disease can spread more quickly.

- State that a disadvantage of using monocultures is that more fertilisers are used as growing the same crop year after year means that the nutrients needed by the crop are used up.

- State that a disadvantage of using monocultures is that soil is compacted by machinery. This reduces soil oxygen, which is needed by plants for respiration.

- State that a disadvantage of using monocultures is that removal of hedgerows reduces habitats available to wild animals.

Fertilisers

Fertilisers are chemicals that are added to soil to improve **plant growth**. Fertilisers have different **minerals** added to improve the plant growth. **Nitrates** and **magnesium** are minerals added to soil in fertilisers (fig 3.86).

Nitrates are important to plant growth. They are needed to produce nucleic acids, amino acids and proteins. Magnesium is important to plant growth, as it is needed in the formation of the pigment chlorophyll.

Chlorophyll is needed by plants to capture light energy needed for photosynthesis.

Fertilisers can **leach** from the soil into nearby **fresh water**. The leaching occurs when rain washes excess fertiliser off the soil. 'Eutrophic' means well nourished. Therefore eutrophication is a term used to describe water enriched with minerals. Eutrophication caused by fertilisers can increase **algal blooms**.

Fig 3.86 *Heavy machinery is used to spray fertilisers on farmland. This can lead to soil becoming compacted*

Algal blooms

Algae are tiny plants that live in water. They need nutrients to grow. When they grow in an uncontrolled manner they can form an algal bloom. Uncontrolled growth occurs in good growing conditions. The conditions needed for growth are a supply of minerals, light and warm temperatures (fig 3.87).

Algal blooms cover the surface of the water and block light reaching green plants under the surface. This prevents the plants gaining light. Plants are unable to carry out photosynthesis. This can lead to a **reduction in oxygen levels** in the water.

Algal blooms can result in the death of deep-water green plants. Bacteria living in the water feed on the dead plant material. The bacteria use up the oxygen in the water. This can lead to a reduction in oxygen levels in the water. The result is that many animals die, leading to a decrease in biodiversity.

Fig 3.87 *Algal blooms prevent light reaching green plants living under the water surface*

Nutrients required for plant growth

Nitrates are important for plant growth. Without nitrates, plant growth would be reduced. The leaves would be yellow instead of green. The plant would not be able to produce enzymes needed to catalyse reactions.

Magnesium is important for plant growth. Without magnesium plant growth would be reduced. The leaves of the plants would appear yellow instead of green. The plant would not be able to photosynthesise efficiently. Crop plants would be unable to produce excess sugar. It is the excess sugar that the plant turns into the useful product, which is harvested.

Reducing problems caused by using fertilisers

Because of the need to provide food for the increasing human population farmers need to use fertilisers. If farmers were to use natural fertilisers such as manure and compost, the risk of algal blooms would be reduced. This is because these natural fertilisers do not dissolve in water as easily and are therefore less likely to be washed into rivers.

Farmers could change to using only natural fertilisers near fresh water. Farmers could also use less fertiliser. Traditional farming methods such as crop rotation used less chemical fertilisers.

In crop rotation, a different crop is grown each year. This is beneficial, as different crops have different mineral needs. Planting crops with deep roots one year and crops with shallow roots the next year gives the soil nutrient levels a chance to recover. This also improves the soil structure.

🏁 Activities

Activity 3.5.4 Working individually

(a) Give a reason why farmers spray fertilisers onto their crops.

(b) Name two minerals that are added to the soil in fertilisers.

(c) Explain how an algal bloom may be caused in fresh water.

(d) Explain why oxygen levels in water may be reduced by an algal bloom.

(e) State the use plants make of nitrates and magnesium.

Activity 3.5.5 Working individually

Six conical flasks were set up to investigate the effect fertilisers had on the growth of algae found in a pond.

A single sample of pond water was collected. The pond water was shaken and then each flask had the same volume of pond water added.

Each flask had a different volume of fertiliser added. The volumes added are shown in the table below.

Flask	Volume of fertiliser added (cm³)	Dry weight of algae produced (g)
1	0	0.05
2	0.5	0.10
3	1.0	0.18
4	1.5	0.20
5	2.0	0.34
6	4.0	0.34

(a) State the variable that was investigated in this experiment.

(b) Give two variables that would have to have been kept constant to make this a valid experiment.

(c) Suggest how the reliability of the experiment could have been improved.

(Continued)

🌳 Biology in context

Algal blooms can have an adverse effect on human health. Toxins produced by the algae can cause allergic skin reactions or contact dermatitis when swimming in contaminated water. Swallowing small volumes of this water can cause headaches, stomach cramps, nausea and diarrhoea.

☄ Make the link – Biology

In Unit 1 Chapter 4 (pages 54–55), you learned of the importance of amino acids in the formation of proteins. Remember that plants need to make proteins. They make protease enzymes to break down proteins into amino acids. They make growth substances such as indole acetic acid, which is important in phototropism.

☄ Make the link – Health and Wellbeing

Organic food is grown without using fertilisers and pesticides. Nutrients needed by plants are supplied by adding manure. Pests are removed by hand. Why is organic food more expensive than intensively farmed food?

🔍 Hint

Phototropism is the movement of plants towards light. The growth substance indole acetic acid causes the cells furthest away from the light to become longer. This causes the plant to grow towards light.

(d) Explain why conical flasks were used instead of beakers.

(e) Suggest why the pond water was shaken before being added to the flasks.

(f) Suggest why the flasks were left on a warm, sunny window ledge.

(g) Explain why one flask had no fertiliser added.

(h) Collect a piece of graph paper. Draw a line graph to show the relationship between the volume of fertiliser added and the dry weight of algae produced.

(i) Describe the relationship between the volume of fertiliser added and the dry weight of algae.

(j) Explain why the dry weight of algae may have levelled off at concentrations above 2 g.

(k) State how many times higher the dry weight of algae produced at 1.5 cm^3 of fertiliser added was than the control.

(l) Suggest why the dry weight of algae was used to compare the results rather than the fresh mass.

Activity 3.5.6 Working individually

Tomato plants are grown around the World. They contain lycopene. Lycopene is an example of a carotene. Carotenes are photosynthetic pigments that are used by plants to capture light energy needed for photosynthesis.

Lycopene is a natural antioxidant. An antioxidant is a substance produced by cells that helps the body fight infections. Eating tomatoes is thought to have health benefits for many organs, including the heart.

Tomato plants are known to require magnesium to grow. Soil lacking magnesium leads to poor growth of tomato plants. Imagine a tomato grower has written to you asking for advice regarding poor growth of tomato plants. You suspect the problem may be caused by a magnesium deficiency. Reply to the grower, offering advice. You should include:

(a) The signs of magnesium deficiency in tomato plants.

(b) The effect of magnesium deficiency on tomato plant growth.

(c) The type of soil that is most likely to lead to magnesium deficiency.

(d) Methods that could be used to treat magnesium deficiency.

(e) Why tomato plants need magnesium for growth.

Your letter should be 150–200 words long.

I can:

- State that a fertiliser is a chemical that improves plant growth.
- State that a fertiliser is added to soil to increase the minerals in the soil.
- State that fertilisers contain nitrates and magnesium.
- State that nitrates are important for plant growth, as they are needed to make nucleic acids and amino acids.
- State that magnesium is needed for the formation of chlorophyll.
- State that fertilisers can be leached from soil and washed into fresh water.
- State that an algal bloom occurs when fresh water becomes enriched with minerals due to leaching.
- State that algal blooms lead to a reduction in oxygen levels in water, as plants are unable to obtain light for photosynthesis and die.
- State that dead plants increase the food available for bacteria, which then multiply and use up even more oxygen.
- State that the use of manure and compost as fertilisers reduces the risk of algal blooms, as they do not dissolve in water easily so are less likely to be washed into freshwater.

Pesticides

Pesticides are chemicals that are sprayed onto **crops** to reduce competition from other organisms (fig 3.88). There are several types of pesticides. Some examples are given below.

1. Herbicides are used to kill plants.
2. Fungicides are used to kill fungal infections.
3. Insecticides are used to kill insects.
4. Bactericides are used to kill bacteria.

Fig 3.88 *Spraying pesticides reduces the effects of competition. They can have a negative effect on human health*

If competition between the crop plant and other organisms is reduced then the crop yield is increased. However, pesticide use has been responsible for a decrease in biodiversity. This is because some pesticides sprayed onto crops **accumulate in the bodies of organisms over time**. This was observed after the use of the insecticide DDT.

Pesticides and the food chain

DDT is a powerful insecticide first used in farming after World War 2. From 1972 it was banned for agricultural use in the United States. It has since been banned worldwide.

The ban was necessary as the chemicals in DDT were found to be non-biodegradable. This means that the chemicals in DDT could not be broken down by the animals that ate them. Therefore every time food contaminated with DDT was eaten, DDT built up in the animal. This is known as **bioaccumulation**.

Bioaccumulation of DDT occurred along the food chains. Organisms at each link in the food chain were found to accumulate DDT in their tissues.

DDT entered the food chain in plants. Consumers of plants took DDT into their bodies. Plants tolerated the DDT due to the low levels of the chemical. However, consumers of plants were found to have more DDT in their bodies than was present in the plants. The level had accumulated. The level in the bodies of consumers continued to accumulate through to the top carnivore in the food chain or web. The top carnivores contained the highest level of DDT (fig 3.89).

DDT was responsible for a large reduction in bird numbers. This was not only due to the increase in toxicity. A large concentration of DDT was found to make it difficult for birds to absorb calcium. It was also thought to alter hormones needed for reproduction. Birds laid eggs with thin shells, and the chicks hatched before they were strong enough to survive.

As certain pesticides pass along **food chains, toxicity increases and can reach lethal levels**.

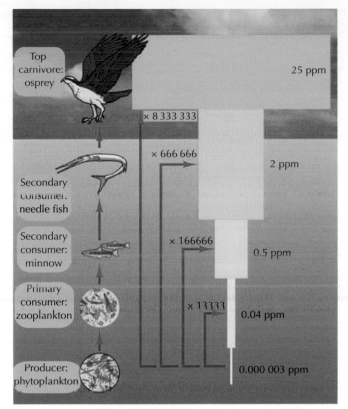

Fig 3.89 *DDT is bioaccumulated in the tissues of organisms as it passes along the food chain*

Biology in context

Ospreys are birds that became extinct in Scotland early in the twentieth century (fig 3.90). They reappeared naturally in 1954. By the 1970s there were still only 14 breeding pairs. The slow increase was in part due to bioaccumulation of DDT. The Osprey Centre at Loch Garten is a conservation site.

Fig 3.90 *DDT bioaccumulated in the tissue of ospreys. This led to thinning of eggshells*

Activities

Activity 3.5.7 Working individually

(a) Explain why pesticides are sprayed onto crops.

(b) Name four different types of pesticides used in farming and state why each one is used.

(c) Give the term used to describe the build-up of pesticides in organisms.

(d) Suggest why DDT accumulated in the organisms at the top of the food chain.

(e) Explain the effect of DDT on reproduction in birds.

Activity 3.5.8 Working individually

The table below shows the average levels of DDT in human body fat of people living in the United States during the period 1942–1982.

Year	DDT level in human body fat (mg/g fat)
1942	0
1950	5.3
1956	10.6
1962	12.6
1963	10.3
1968	12.5
1970	11.6
1972	9.2
1974	6.7
1976	5.5
1978	4.8
1980	0
1982	0

(a) Suggest a reason for the lack of DDT in human body fat in 1942.

(b) State the number of years it took for the DDT level in human body fat to double after it was first detected.

(c) Suggest a reason for the lack of DDT from 1980 onwards.

(d) DDT is an insecticide. Suggest a reason why it might have been found in human body fat.

- After the initial setup costs it may be much cheaper to control pests using biological control.

- There are no problems of chemicals contaminating the environment or the food chain.

- Most pests do not become resistant to the treatment, as no chemicals are used. The same method of control can be used year after year.

Disadvantages of biological control

- Its use in small confined areas, e.g. greenhouses, has found more success than in fields. It is not as easy for the predators to leave the environment.

- It can initially be very expensive to practise the technique due to the high development costs.

- Some introduced predators have become pests themselves. This happened with the cane toad (fig 3.93).

- Not all the pests will be killed. Some remain and still cause damage. Even with biological control, pesticides may still need to be used.

Fig 3.93 *Cane toads were introduced in Australia to control cane beetles. There were no natural predators so the numbers increased dramatically. The toads ate non-harmful organisms. They have become a pest species themselves*

Biological control in action

An early example of biological control occurred when the virus *Myxoma* was released into rabbit populations. The virus caused myxomatosis. Rabbits infected with the virus die a week after infection. The rabbits develop sores to their eyes and become blind.

The virus killed many rabbits, so crop damage was reduced. An unexpected result was that rabbits became immune to the virus and it became an unsuccessful method of control.

Ladybirds eat a variety of insects (fig 3.94). They have been introduced to kill aphids and scale insects. Aphids feed on the sap of plants. They can introduce pathogens into plants. Plants infested with aphids tend to produce few, small flowers. Crop plants need to produce flowers to produce the useful product. It is thought that one ladybird is able to eat up to 50 greenfly in a day and 5000 over its life.

Scale insects also feed on the sap of plants. Some varieties leave a sticky substance on leaves. This encourages black mould to grow. The mould damages the leaves of the crop plant.

Fig 3.94 *Ladybirds are the natural predators of aphids*

🌱 Biology in context

Biological control can also be used to control weeds. A weed is a plant that grows out of control. This is the case with Japanese knotweed. In 2010 a trial was undertaken where an insect native to Japan was released into selected sites in England. It was hoped that the insect would feed on the sap of the weed and stunt its growth. If it is successful it could save the UK economy more than £150 million a year.

🔍 Hint

GM crops and biological control are used as alternatives to pesticides. These technologies cost less and are potentially less damaging to the environment and human health.

⚗️ Make the link — Biology

You studied food chains and food webs in Unit 3 Chapter 17 (pages **258–262**). Remember if a predator is a successful hunter, the numbers of prey decline. The prey is the pest species. For this reason biological control should not become a problem so long as the predator is carefully chosen.

⚗️ Make the link — Religious and moral education

You may have studied ethics. Ethics involves deciding the rights and wrongs of situations. Can you apply your knowledge of ethics to breeding organisms to control the growth of other organisms?

🔵 Activities

Activity 3.5.14 Working individually

(a) State why farmers use biological control in farming.

(b) Explain how biological control can reduce the use of chemicals in farming.

(c) Give three advantages of using biological control rather than pesticides.

(d) Give three disadvantages of using biological control rather than pesticides.

(e) Name two organisms that have been used in biological control in farming.

Activity 3.5.15 Working individually

The caterpillar moth (*Cactoblastis*) has been introduced to kill cacti (*Opuntia*).

(a) Look up the following web address:

http://www.northwestweeds.nsw.gov.au/cactoblastis.htm

(b) Produce a newspaper article of around 200–300 words informing readers of the success in using *Cactoblastis* in biological control. Your article should be one A4 page in size.

Include the following information:

- The natural habitat of the moth.
- The plant that it has been used to control. Explain why the use of the moth has been effective.
- A photograph of both the moth and the plant.
- A description of the moth that would allow it to be recognised.
- A description of the life cycle of the moth including information on the following stages – eggsticks, larvae and moths.
- Natural predators of the moth.

I can:

- State that intensive farming methods have been shown to have a harmful effect on the environment.
- State that biological control uses natural predators to control pests in farming.
- State that ladybirds and caterpillar moths have been used in biological control.
- State that biological control reduces the effects of chemical pesticides that may kill helpful insects that act as pollinators.
- State that biological control reduces the effects of chemical pesticides, as the pests do not usually become resistant to the treatment.
- State that biological control reduces the effects of chemical pesticides, as bioaccumulation does not occur, and pesticides do not enter the food chain.
- State that biological control reduces the effects of chemical pesticides. Since no chemicals are sprayed there is no risk to human health.
- State that the introduced predator is specific to the pest; therefore biodiversity is not reduced.
- State that biological control is more effective in small growing areas, since the predator is less likely to move out of the area.
- State that in some cases the natural predator has become a pest itself, e.g. cane toad.

Unit 3– Assessment

Section A

1. Two examples of biotic factors are

   A   grazing and wind speed

   B   grazing and predation

   C   predation wind speed

   D   predation and humidity

2. At each level in a food chain, energy is lost as

   A   heat only

   B   movement only

   C   heat and movement

   D   heat, movement and undigested materials

3. Competition between organisms is most intense when it is

   A   interspecific competition between members of the same species

   B   intraspecific competition between members of the same species

   C   interspecific competition between members of different species

   D   intraspecific competition between members of different species

4. Mutations are random changes to an organism's genetic material.

   Mutations are

   A   advantageous only

   B   disadvantageous only

   C   neutral only

   D   the only source of new alleles

5. The numbers of the dark variety of peppered moths increased in Britain in industrial areas during the industrial revolution.

   The increase was due to

   A   increasing rates of mutation of the dark form

   B   dark moths moving into the area

   C   predators selecting more of the light-coloured moths

   D   light moths dying due to disease

6. Speciation describes the evolution of new species. A new species is considered to have evolved when a population

   A   has members that can no longer produce fertile offspring

   B   can no longer interbreed with the original population

   C   shows increased variation due to mutations

   D   shows decreased variation due to lack of mutations

7. Indicator species indicate levels of pollution by their presence or absence. Mayfly nymphs indicate unpolluted water. The table below shows the results of a survey carried out on four rivers.

Mayfly nymphs would be found in river ……… .

River	Quantity of organic matter in water	Oxygen level	Number of bacteria in water
A	Low	Low	Low
B	Low	High	Low
C	High	High	High
D	High	Low	High

8. Information about fertiliser usage in Scotland is shown in the graph below.

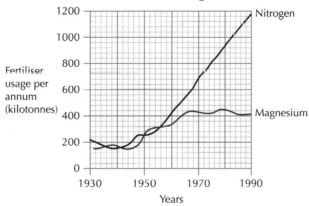

Which of the following statements about magnesium usage is correct?

   A   Magnesium usage per year reached 1200 kilotonnes.

   B   Magnesium usage per year never rose above 420 kilotonnes.

   C   Maximum magnesium usage per year was 440 kilotonnes.

   D   Magnesium usage never exceeded nitrogen usage.

9. The graph below shows the numbers of foxes and rabbits found in a forest over a ten-week period. Rabbits are the prey of foxes.

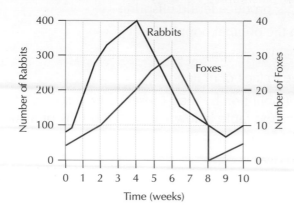

The ratio of foxes to rabbits in week 2 is

A   1:3

B   30:1

C   1:30

D   10:300

10. The graph below shows the average height of trees in a forest over a period of 25 years.

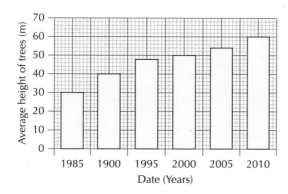

The percentage increase in the average height of trees between 1990 and 2000 was

A   10%

B   20%

C   25%

D   45%

Section B

1. The diagram below shows part of a woodland food web.

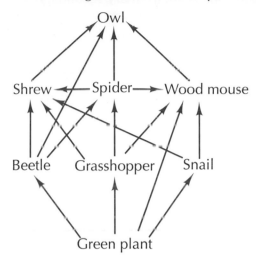

(a) Use five organisms from the food web to construct a food web below.

_____ → _____ → _____ → _____ →

_____

1

(b) Decide if each of the following statements is **True** or **False**, and tick (✓) the appropriate box.

If the statement is **False**, write the correct word in the **correction box** to replace the word(s) underlined in the statement.

Statement	True	False	Correction
In this food web, grasshoppers are <u>carnivores</u>.			
The population of green plants have the <u>highest</u> biomass.			
In this food web, the wood mouse passes its energy to <u>spiders</u>.			

3

(c) Explain why a decrease in shrew numbers may lead to an increase in wood mouse numbers.

_____

_____

1

(d) Underline **one** option in each set of brackets to make the following sentence correct.

1

Beetles are the (predators/prey) of spiders. Snails are (primary/secondary) consumers.

(e) Give the term used to describe the role an organism plays within an ecosystem.

_____

1

2. In Scottish grasslands, sheep are often found as grazers. A small flock of sheep was introduced into a large area of ungrazed grassland.

   (a) Explain why this would increase biodiversity within this area.

   _____

   _____

2

   (b) In another area in Scotland, a grassland was in danger of being overgrazed by rabbits. To control this, a disease was introduced that killed rabbits but did not harm other species.

   Give the name of this technique, which is designed to reduce rabbit numbers without adding chemicals to the environment.

   _____

1

3. The light intensity inside and outside a woodland was measured.

   The results are shown in the table below.

Month	Average daily light intensity (units)	
	Outside woodland	Inside woodland
January	15	14
February	17	16
March	22	19
April	26	23
May	29	26
June	33	18
July	36	10
August	31	10
September	27	10
October	22	10
November	18	16
December	15	14

(a) Explain why the light intensities outside and inside the woodland showed the greatest difference between June and October.

_____

_____ 1

(b) State **one** further factor that could have been measured to compare the effect of abiotic factors on the distribution of organisms within and outside the woodland.

Give a possible source of error when measuring the abiotic factor you have **named** and explain how the effect of the error could be minimised.

Abiotic factor _____

Source of error _____ 1

Method of minimising error _____

_____ 1

(c) Quadrats and pitfall traps could be used to sample organisms living in the woodland.

For either quadrats **or** pitfall traps give a possible source of error when using the technique and explain how the effect of the error could be minimised.

Source of error _____

Method of minimising error _____

_____ 1

4. An investigation was carried out into the effect of competition on the survival of grass seedlings. Five identical petri dishes were set up as shown in the diagram. Each dish contained a different number of grass seeds as shown in the table below.

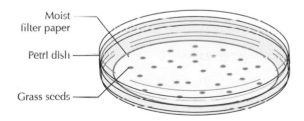

Moist filter paper

Petri dish

Grass seeds

Each dish was left to grow for seven days. The results of the investigation are shown in the table below.

Dish	Number of seeds sown	Number of surviving seedlings after 7 days	Percentage of surviving seedlings after 7 days (%)
1	10	10	100
2	20	18	90
3	40	32	80
4	80	48	60
5	100	35	35

(a) State the variable that was altered in the investigation.

_____ 1

(b) State **two** variables, not already mentioned, that need to be kept the same in this investigation.

1. _____

2. _____ 1

(c) Explain why the percentage of surviving seedlings after 7 days was calculated.

_____

_____ 1

(d) Give one requirement that the grass seedlings may be competing for.

_____ 1

(e) Draw a **line graph** to show the number of seeds planted against the percentage of seedlings surviving after 7 days.

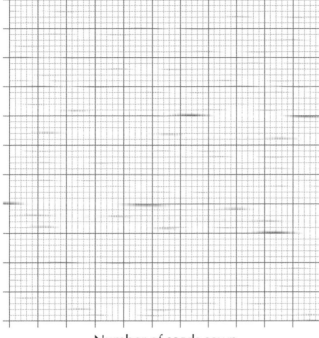

Number of seeds sown

2

5. A mutation is a random change to genetic material. Mutations can be caused by environmental factors such as chemicals.

(a) Give **one other** environmental factor that can cause mutations.

_____ 1

(b) Give **two** factors, other than mutations, that are needed for a new species to evolve during the process of speciation.

1. _____

2. _____ 2

6 An investigation was carried out into the effect of temperature on the growth of young fish. The eggs were transferred to their incubation temperature immediately after fertilisation. At this stage the eggs were 3 mm in length.

The graph below shows the average lengths of the fish 4 weeks after hatching.

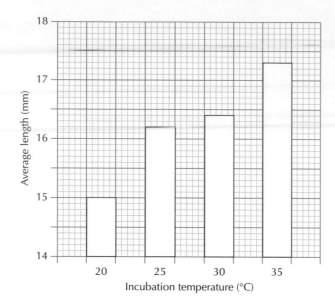

(a) Describe the relationship that exists between the incubation temperature and the average length of the young fish.

_____

_____ 1

(b) State the conclusion that can be drawn from these results.

_____

_____ 1

(c) Predict the average length of the fish if the incubation temperature had been increased to 55°C. Explain your answer.

Predicted length _____ mm   1

Explanation_____

_____ 1

7. Fertilisers and pesticides are chemicals that are used in intensive farming of crop plants. For either fertilisers **or** pesticides, state why the chemical is used **and** describe the effect it has on the **environment**.

_____

_____

_____

_____

_____

_____ 3